Bjørn Kjos-Hanssen

Automatic Complexity

De Gruyter Series in Logic and Its Applications

Volume 12

Bjørn Kjos-Hanssen

Automatic Complexity

A Computable Measure of Irregularity

DE GRUYTER

Mathematics Subject Classification 2020
68R15, 68Q30, 94A55

Author
Prof. Bjørn Kjos-Hanssen
University of Hawaii at Manoa
Department of Mathematics
2565 McCarthy Mall
Honolulu, HI 96822
USA
bjoern.kjos-hanssen@hawaii.edu

ISBN 978-3-11-077481-8
e-ISBN (PDF) 978-3-11-077487-0
e-ISBN (EPUB) 978-3-11-077490-0
ISSN 1438-1893

Library of Congress Control Number: 2023920875

Bibliographic information published by the Deutsche Nationalbibliothek
The Deutsche Nationalbibliothek lists this publication in the Deutsche Nationalbibliografie; detailed bibliographic data are available on the Internet at http://dnb.dnb.de.

Typesetting: VTeX UAB, Lithuania
Printing and binding: CPI books GmbH, Leck

www.degruyter.com

In memory of my father, Odd

Preface

As the 1968 film *2001: A Space Odyssey* gave an enigmatic and scientifically accurate depiction of space flight, so Jeffrey O. Shallit and Ming-Wei Wang's paper *Automatic complexity of strings* [69] from 2001 described what we can call a "state odyssey": journeys through the states of a finite automaton that held the promise of further deep exploration.

While Kolmogorov complexity is only defined "up to an additive constant," automatic complexity gives concrete values. When I start up the Complexity Guessing Game at http://math.hawaii.edu/wordpress/bjoern/software/web/complexity-guessing-game/, I am presented with the string $x = 011000111001010$ of length 15 and asked to guess its complexity, from 1 to 8. As we shall see in this book, in particular in Chapter 5, the best bet is to choose the largest complexity offered (8 in this case) unless you spot something very special about x. This is correct for our x. Next I am asked about the length-25 word

$$y = 1011001111101111010100101 \tag{1}$$

for a guess for its complexity, from 1 to 13. Again I choose the maximum, but in this case, the game responds that the complexity is 12 and that there is a complexity *deficiency* of 1.

The idea of complexity (or randomness) deficiency comes from the study of Kolmogorov complexity. However, automatic complexity is a more manageable (and computable) measure of irregularity. When we say "irregularity" here it is partly a pun: automatic complexity is based on finite automata, which accept regular languages, so irregularity indicates a failure of small finite automata to uniquely identify the string in a sense.

The game provides the finite automaton for y from Equation (1) shown in Figure 1. The reader can check that there is one and only one way to travel through this automaton, starting at q_0 and also ending at q_0 while reading the word y. This uniqueness comes from the unique solution of the constrained Diophantine equation in (2) over $\mathbb{N}$:

$$\begin{gathered} 3w + 6x + 6y + 5z = 25, \\ z > 0 \implies y > 0, \quad y > 0 \implies x > 0, \quad x > 0 \implies w > 0. \end{gathered} \tag{2}$$

This research area was started by Jeff Shallit and Ming-Wei Wang in 2001 [69]. I independently made the same definition in 2009 while teaching the class Math 301 (Discrete Mathematics) at the University of Hawaiï. Subsequently, I wrote several papers which are treated in this book. Moreover, Jordon and Moser wrote a paper on the topic in 2021 [35]. It is my hope that this book will stimulate further work in this area.

As a research tool, and incidentally as a method of cheating at the *Complexity Guessing Game*, I have created a web service to find the complexity of a given word and an illustration of an automaton used in the associated proof [45].

https://doi.org/10.1515/9783110774870-201

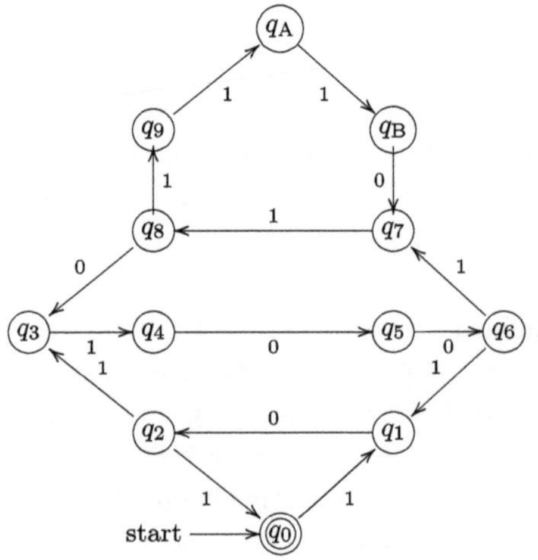

Figure 1: An automaton provided by the *Complexity Guessing Game*.

The Complexity Option Game [46] is a variation on the same idea, inviting the player to implement an exercise policy for a complexity-based financial option. These games include graphical displays of millions of the relevant automata.

How to use this book

Chapter 1, Chapter 2, Chapter 3, and Chapter 4 answer the question "What?" by introducing the basics of state-counting and edge-counting automatic complexity. Chapter 5 answers the question "How?" (Do we work with automatic complexity?), answering a question of Shallit and Wang by estimating the complexity of random words. Chapter 6 and Chapter 7 are an attempt to answer the question "Why?" Conditional automatic complexity gives a perspective on the length-conditional aspect of automatic complexity, and automatic complexity turns out to give an answer to the question what logical depth looks like in practice.

Each chapter contains exercises, most of which come with solutions in the proof assistant Lean. Open research problems also appear in dedicated sections.

While I have tried to keep this book *akamai*, errors may occur and are my responsibility. I would be grateful to receive reports at bjoernkh+acmoi@hawaii.edu.

Honolulu, December 2023

Acknowledgments

I am grateful to many people.

- Andrew J. I. Jones, Dag Normann, Theodore A. Slaman, and many others mentored me in computability and logic.
- In a Discrete Mathematics class in Spring 2009, students Jason Axelson, Chris Ho, and Aaron Kondo wrote a C program that calculated the complexity of all strings of length at most 7. The dedication they put into that program fueled my interest in automatic complexity.
- Logan Axon wrote the first Python script for automatic complexity.
- Kayleigh K. Hyde's 2013 Master's project and her proof of the sharp upper bound for nondeterministic automatic complexity sparked my interest in proving theorems in this area.
- Students who have worked with me on automatic complexity include Samuel D. Birns, Calvin K. Bannister, Swarnalakshmi (Janani) Lakshmanan, and Daylan K. Yogi.
- Jeff Shallit, Achilles Beros, Nikolai Vereshchagin, Sasha Shen, André Nies, Frank Stephan, and Angeliki Koutsoukou-Argyraki provided encouragement and interesting discussions.

I thank De Gruyter staff for their patience throughout this project. This research was supported in part by a grant from Decision Research Corporation (University of Hawaiï Foundation Account #129-4770-4). This work was partially supported by a grant from the Simons Foundation (#704836 to Bjørn Kjos-Hanssen).

https://doi.org/10.1515/9783110774870-202

Contents

1 First steps in automatic complexity

The Kolmogorov complexity of a finite word w is, roughly speaking, the length of the shortest description w^* of w in a fixed formal language. The description w^* can be thought of as an optimally compressed version of w. Motivated by the noncomputability of Kolmogorov complexity, Shallit and Wang [69] studied a deterministic finite automaton analogue.

Their notion of automatic complexity is an automata-based and length-conditional analogue of Sipser's distinguishing complexity *CD* ([73], [53, Definition 7.1.4]). This was pointed out by Mia Minnes in a review in the *Bulletin of Symbolic Logic* from 2012. Another precursor is the length-conditional Kolmogorov complexity [53, Definition 2.2.2].

The automatic complexity of Shallit and Wang is the minimal number of states of an automaton accepting only a given word among its equal-length peers. Finding such an automaton is analogous to the protein folding problem where one looks for a minimum-energy configuration. The protein folding problem may be NP-complete [15], depending on how one formalizes it as a mathematical problem. For automatic complexity, the computational complexity is not known, but a certain generalization to equivalence relations gives an NP-complete decision problem [37].

In this chapter we start to develop the properties of automatic complexity.

1.1 Words

The set of all natural numbers is $\mathbb{N} = \{0, 1, 2, \dots\}$. Following the von Neumann convention, each natural number $n \in \mathbb{N}$ is considered to be the set of its predecessors:

$$0 = \emptyset, \quad 1 = \{0\}, \quad \text{and in general} \quad n = \{0, 1, \dots, n-1\}.$$

The power set $\mathcal{P}(X)$ of a set X is the set of all the subsets of X:

$$\mathcal{P}(X) = \{A \mid A \subseteq X\}.$$

If Σ is an *alphabet* (a set), a *word* (sometimes called *string*) is a sequence of elements of Σ.

For computer implementations it is often convenient to use an alphabet that is an interval $[0, b)$ in $\mathbb{N}$. When we want to emphasize that 0, 1, etc., are playing the role of symbols rather than numbers, we often typeset them as $\mathtt{0}$, $\mathtt{1}$, etc., respectively.

We denote concatenation of words by $x +\!\!+\, y$ or by juxtaposition xy. In the word $w = xyz$, y is called a *subword* or *factor* of w. Infinite words are denoted in boldface. For example, there is a unique infinite word $\mathbf{w}$ such that $\mathbf{w} = 0\mathbf{w}$, and we write $\mathbf{w} = 0^\infty$.

Let $\preceq$ denote the prefix relation, so that $a \preceq b$ iff a is a prefix of b, iff there is a word c such that $b = ac$.

https://doi.org/10.1515/9783110774870-001

The concatenation of a word x and a symbol a is written $x :; a$ if a is appended on the right and $a :: x$ if a is appended on the left. It may seem most natural to define Σ^* by induction using $x :; a$, at least for speakers of languages where one reads from left to right. The Lean proof assistant [16] (version 3) uses $a :: x$. That approach fits well with co-induction, if infinite words are ordered in order type $\mathbb{N}$ [50].

For $n \in \mathbb{N}$, Σ^n is the set of words of length n over Σ. We may view $\sigma \in \Sigma^n$ as a function with domain n and range Σ.

We view functions $f : A \to B$ as subsets of the Cartesian product $A \times B$,

$$f = \{(x,y) \mid y = f(x)\}.$$

The set Σ^n is both the set of functions from n to Σ and the Cartesian product $(\Sigma^{n-1}) \times \Sigma$ if $n > 0$. The empty word is denoted ε and the first symbol in a word x is denoted $x(0)$.

We can define $\Sigma^* = \bigcup_{n\in\mathbb{N}} \Sigma^n$. More properly, the set Σ^* is defined recursively by the rule that $\varepsilon \in \Sigma^*$, and whenever $s \in \Sigma^*$ and $a \in \Sigma$, then $s :; a \in \Sigma^*$. We define *concatenation* by structural induction: for $s, t \in \Sigma^*$,

$$t \mathbin{++} \varepsilon = t,$$
$$t \mathbin{++} (s :; a) = (t \mathbin{++} s) :; a.$$

Definition 1.1. The *length* of a word $s \in \Sigma^*$ is defined by induction:

$$|\varepsilon| = 0,$$
$$|s :; a| = |s| + 1.$$

If $A \subseteq B$ and $f \subseteq B \times C$, then $f \upharpoonright A = \{(x,y) \in f \mid x \in A\}$ is the restriction of f to A.

The word σ is also denoted $\langle \sigma(0), \sigma(1), \ldots, \sigma(|\sigma| - 1)\rangle$. By convention, instead of $\langle 0, 1, 0\rangle$ we write simply $\mathtt{010}$.

Example 1.2. We have

$$\begin{aligned} \langle \mathtt{0}\rangle \mathbin{++} \langle 1, \mathtt{0}\rangle &= \langle \mathtt{0}, 1, \mathtt{0}\rangle \\ &= \mathtt{0} :: \langle 1, 0\rangle \\ &= \langle 0, 1\rangle :; \langle 0\rangle. \end{aligned}$$

1.1.1 Occurrences and powers

In this subsection we state some results from the subject "combinatorics on words" that will be used frequently.

The statement occurs(x, k, y) that x occurs in position k within y may be defined by induction on n:

$$\begin{aligned}\text{occurs}(x,0,y) &\iff \exists z,\ x +\!\!+\, z = y,\\ \text{occurs}(x,n+1,y) &\iff \exists a \in \Sigma,\ \text{occurs}(a :: x, n, y).\end{aligned}$$

Example 1.3. We have occurs(na, 2, banana) and occurs(na, 4, banana).

The number of occurrences can be defined, without defining a notion of "occurrence," as the cardinality of $\{k \in \mathbb{N} : \text{occurs}(x,k,y)\}$. To define disjoint occurrences, we should have a notion of "occurrence."

Definition 1.4. Two occurrences of words a (starting at position i) and b (starting at position j) in a word x are *disjoint* if $x = uavbw$, where u, v, w are words and $|u| = i$, $|uav| = j$.

Type-theoretically [14] we may say that an occurrence of x in y is a pair (k,h), where h is a proof that occurs(x,k,y). Of course, we could also say that the occurrence is simply the number k, or even the triple (x,k,y), but in that case the object does not have its defining property within it, so to speak: the number k and the triple (x,k,y) exist even when x does not occur at position k in y.

To naturally speak of disjoint occurrences we provide Definition 1.5. Note that we primarily use zero-based words in this book, i. e., other things being equal we prefer to call the first letter of a word x_0, rather than x_1.

Definition 1.5. A word $x = x_0 \dots x_{n-1}$, or an infinite word $x = x_0x_1\dots$, with each $x_i \in \Sigma$, is viewed as a function $f_x : n \to \Sigma$ with $f_x(i) = x_i$. An *occurrence* of $y = y_0 \dots y_{m-1}$ in x is a function $f_x \restriction [a, a+m-1]$ such that $f_x(a+i) = y_i$ for each $0 \le i < m$.

For $a, b \in \mathbb{N}$, let $[a,b] = \{x \in \mathbb{N} \mid a \le x \le b\}$. Two occurrences $f_x \restriction [a,b], f_x \restriction [c,d]$ are *disjoint* if $[a,b] \cap [c,d] = \emptyset$.

If moreover $[a,b+1] \cap [c,d] = \emptyset$, then the occurrences are *strongly disjoint*.

A subword that occurs at least twice in a word w is a *repeated* subword of w. Let $k \in \mathbb{N}, k \ge 1$. A word $x = x_0 \dots x_{n-1}, x_i \in \Sigma$, is *k-rainbow* if it has no repeated subword of length k: there are no $0 \le i < j < n-k$ with $x \restriction [i, i+k-1] = x \restriction [j, j+k-1]$. A 1-rainbow word is also known simply as *rainbow*.

In particular, the empty word ε occurs everywhere in every word. However, each word has exactly one occurrence of ε, since all empty functions are considered to be equal (in set theory, at any rate).

Having properly defined "occurrence" in Definition 1.5, we can state and prove the trivial Lemma 1.6.

Lemma 1.6. *Suppose n, t are positive integers with $t \le n+1$. A word of length n has $n+1-t$ occurrences of subwords of length t.*

Proof. Let x be a word of length n. The occurrences of subwords of length t are

$$f_x \restriction [0, t-1], f_x \restriction [1, t], \dots, f_x \restriction [n-t, n-t+(t-1)].$$ □

Theorem 1.7. *Let $k, t \in \mathbb{N}$ with $k \geq 1$. In an alphabet of cardinality k, a t-rainbow word has length at most $k^t + t - 1$.*

Proof. There are k^t words of length t. Thus, by Lemma 1.6, for a t-rainbow word of length n we have $n + 1 - t \leq k^t$. □

Theorem 1.7 has a converse, Theorem 1.9. To prove it we shall require the notion of a de Bruijn word.

Definition 1.8. A *de Bruijn word of order n* over an alphabet Σ is a sequence y such that every $x \in \Sigma^n$ occurs exactly once as a cyclic substring of y.

Theorem 1.9. *Let $k, t \in \mathbb{N}$ with $k \geq 1$. In an alphabet of cardinality k, there exists a t-rainbow word of length $k^t + t - 1$.*

Proof. Case $t = 0$: The empty word is a 0-rainbow word of length 0. Case $t = 1$: A 1-rainbow word of length k exists, namely, any permutation of the symbols in Σ (Definition 1.5). For $t > 1$, let x be a *de Bruijn* word $B(k, t)$ of length k^t and let $w = x^2 \upharpoonright (k^t + t - 1)$. □

Definition 1.10. Let a be a word of length n and let a_i be the ith letter of a for $1 \leq i \leq n$. We define the uth power of a for certain values of $u \in \mathbb{Q}_{\geq 0}$ (the set of nonnegative rational numbers) as follows. For $u \in \mathbb{N}$:

$$a^0 = \varepsilon,$$
$$a^{n+1} = a^n ++ a.$$

- If $u = v + k/n$, where $0 < k < n$ and k is an integer, then a^u denotes $a^v a_1 \ldots a_k$ and is called a u-power.

The word w is u-power-free if no nonempty v-power, $v \geq u$, occurs (Definition 1.5) in w. In particular, 2-power-free is called *square-free* and 3-power-free is called *cube-free*. Let $\mathbf{w}$ be an infinite word over the alphabet Σ and let x be a finite word over Σ. Let $u > 0$ be a rational number. The word x is said to occur in $\mathbf{w}$ with exponent u if x^u occurs in $\mathbf{w}$ (Definition 1.5).

The reader may note that the definition of u-power-free is perhaps not obvious. For example, the word *aba* is not 1.49-power-free: while it contains no 1.49-power, it contains a 1.5-power. This way of defining things enables Theorem 1.12 and goes back at least to Krieger [51, page 71].

Definition 1.11. The *critical exponent* $\mathrm{ce}(w)$ of an infinite word $\mathbf{w}$ is defined by

$$\mathrm{ce}(w) = \sup\{\alpha \in \mathbb{Q} \mid \mathbf{w} \text{ contains some } \alpha\text{-power}\}.$$

Theorem 1.12 (Krieger [51]). *The critical exponent of* $\mathbf{w}$ *is equal to*

$$\inf\{\alpha \in \mathbb{Q} \mid \mathbf{w} \textit{ is } \alpha\textit{-power-free}\}.$$

Proof. Let

$$S = \{\alpha \in \mathbb{Q} \mid \mathbf{w} \text{ is } \alpha\text{-power-free}\},$$
$$T = \{\alpha \in \mathbb{Q} \mid \mathbf{w} \text{ contains some } \alpha\text{-power}\}.$$

Paying careful attention to Definition 1.10, the word $\mathbf{w}$ is α-power-free iff for all $\beta \geq \alpha$, $\mathbf{w}$ contains no β-power. Therefore, S is upward closed. On the other hand, T is an upward dense subset of the complement of S. Therefore, $\mathrm{ce}(w) = \sup T = \inf S$. □

As an example of Definition 1.10, we have $\mathtt{0110}^{3/2} = \mathtt{011001}$. Note that the expected power rule for exponents fails for word exponentiation. In general, $(x^a)^b \neq x^{ab}$, for instance

$$(01)^3 = 010101 \neq 010010 = ((01)^{3/2})^2.$$

Fix an alphabet Σ and let Σ^+ denote the set of nonempty words over Σ.

Lemma 1.13 (Lyndon and Schützenberger [56]; see [67, Theorem 2.3.2]). *Let* $c, d, e \in \Sigma^+$. *The equation* $cd = de$ *holds iff there exist a nonempty word* u*, a word* v*, and a natural number* p *such that* $c = uv$*,* $d = (uv)^p u$*, and* $e = vu$*.*

Theorem 1.14 (Lyndon and Schützenberger [56]; see [67, Theorem 2.3.3]). *Let* $x, y \in \Sigma^+$. *Then the following four conditions are equivalent:*

1. $xy = yx$;
2. *there exist* $z \in \Sigma$ *and integers* $k, l > 0$ *such that* $x = z^k$ *and* $y = z^l$;
3. *there exist integers* $i, j > 0$ *such that* $x^i = y^j$.

1.2 Automata

In Definition 1.15, some basic objects of study in this book are introduced.

Definition 1.15. Let Σ be a finite alphabet and let Q be a finite set whose elements are called *states*. A *nondeterministic finite automaton* (NFA) is a 5-tuple $M = (Q, \Sigma, \delta, q_0, F)$. The *transition function* $\delta : Q \times \Sigma \to \mathcal{P}(Q)$ maps each $(q, b) \in Q \times \Sigma$ to a subset of Q. Within Q we find the *initial state* $q_0 \in Q$ and the set of *final states* $F \subseteq Q$. The function δ is extended to a function $\delta^* : Q \times \Sigma^* \to \mathcal{P}(Q)$ by structural induction:

$$\delta^*(q, \varepsilon) = \{q\},$$
$$\delta^*(q, \sigma :: i) = \bigcup_{s \in \delta^*(q,\sigma)} \delta(s, i).$$

```
def deterministic (self, t):
    return not any(
        (p, p1, b) in t and (p, p2, b) in t and p1 != p2
        for p  in range(0, self.q) for p1 in range(0, self.q)
        for p2 in range(0, self.q) for b  in range(0, self.b)
    )
```

Figure 1.1: Checking for determinism of a set of triples (p1 , p2 , b), where $p_2 \in \delta(p_1, b)$.

Overloading notation, we may also write $\delta = \delta^*$. The *language accepted by M* is

$$L(M) = \{x \in \Sigma^* \mid \delta(q, x) \cap F \neq \emptyset\}.$$

A *deterministic finite automaton* (DFA) is also a 5-tuple $M = (Q, \Sigma, \delta, q_0, F)$. In this case, $\delta : Q \times \Sigma \to Q$ is a total function and is extended to $\delta^* : Q \times \Sigma^* \to Q$ by

$$\delta^*(q, \sigma :; i) = \delta(\delta^*(q, \sigma), i). \tag{1.1}$$

If the domain of δ is a subset of $Q \times \Sigma$, M is an *incomplete DFA*. In this case, M coincides with an NFA with a special property which we can check for effectively (Figure 1.1). We denote a partial function f from A to B by $f : (\subseteq A) \to B$ or $f \overset{\cdot}{:} A \to B$.

Finally, the set of words accepted by M is

$$L(M) = \{x \in \Sigma^* \mid \delta(q, x) \in F\}.$$

Automata may be viewed as an instance of graph theory, where the automaton is a directed graph (digraph) with labeled edges, a state is a vertex, and a transition is an edge. This point of view we introduce now, and return to frequently.

Definition 1.16. A *digraph* $D = (V, E)$ consists of a set of *vertices* V and a set of *edges* $E \subseteq V^2$. Let $s, t \in V$. Let $n \geq 0$, $n \in \mathbb{Z}$. A *walk* of length n from s to t is a function $\Delta : \{0, 1, \ldots, n\} \to V$ such that $\Delta(0) = s$, $\Delta(n) = t$, and $(\Delta(k), \Delta(k+1)) \in E$ for each $0 \leq k < n$.

A *cycle* of length $n = |\Delta| \geq 1$ in D is a walk from s to s, for some $s \in V$, such that $\Delta(t_1) = \Delta(t_2)$, $t_1 \neq t_2 \implies \{t_1, t_2\} = \{0, n\}$. Two cycles are *disjoint* if their ranges are disjoint.

Equation (1.1) may be viewed as a closure property: if $(q, \sigma, r) \in \delta^*$, then $(q, \sigma :; i, \delta(r, i)) \in \delta^*$. Formally, then, δ^* is the intersection of all functions that contain δ (viewing symbols as length-1 strings) and is closed under this closure property. Using the fact that $\mathbb{N}$ is well ordered and is the range of the length function on strings, we can define $G(n)$ to be δ^* restricted to words of length n, and then $G(n)$ is defined from $G \upharpoonright n$ by $G(n) = F(n, G \upharpoonright n)$, where

$$F(n,g) = \begin{cases} \delta, & n = 0, \\ \{(q, \sigma :: i, \delta(r,i)) \mid (q,\sigma,r) \in g(n-1)\}, & n > 0, n-1 \in \mathrm{dom}(g), \\ \emptyset, & n > 0, n-1 \notin \mathrm{dom}(g). \end{cases}$$

Note that $(G \restriction n)(n-1) = G(n-1)$ so that the third clause does not occur in our application. Since $G \restriction 0$ is the empty function, we need to carve out the special case $n = 0$.

Theorem 1.17 is the special case of the wellorder $(\mathbb{N}, <)$ from Schimmerling [66, Theorem 3.8]. We apply it with $B = \{f \mid f \vdots Q \times \Sigma^* \to Q\}$.

Theorem 1.17. *Let B be a set and let P be the set of all partial functions from $\mathbb{N}$ to B. For each $F : \mathbb{N} \times P \to B$, there is a unique function $G : \mathbb{N} \to B$ such that*

$$G(n) = F(n, G \restriction n)$$

for all $n \in \mathbb{N}$.

Remark 1.18. Sipser, 2nd edition [74] does not introduce δ^* at all but discusses everything in terms of a sequence of values of δ. Shallit [67] and Hopcroft and Ullman [28] introduce δ^* but do not give a justification for its existence.

Definition 1.19. Let Σ be an alphabet. Let DFA_Σ denote the class of all DFAs over Σ and let $\mathrm{partDFA}_\Sigma$ denote the class of all partial DFAs over Σ. If a DFA M over Σ has the property that it accepts a string x, but no other strings of length $|x|$, i. e.,

$$L(M) \cap \Sigma^{|x|} = \{x\},$$

then we say that M *accepts x uniquely*. Let $x \in \Sigma^*$ with $|x| = n$. Define $A(x, \Sigma)$, the *automatic complexity* of x (with respect to Σ), to be the least number of states in any $M \in \mathrm{DFA}_\Sigma$ such that M accepts x uniquely.

If we consider x to be a function $x : [n] \to \Sigma$, then $x \in \Sigma^*$, where $\Sigma = \mathrm{range}(x)$. We define $\tilde{A}(x) = A(x, \mathrm{range}(x))$.

Similarly, we define $A^-(x, \Sigma)$, the *partial automatic complexity* of x, to be the least number of states in any $M \in \mathrm{partDFA}_\Sigma$ such that $L(M) \cap \Sigma^n = \{x\}$.

In most cases below, if we write $A(x)$ we intend $\tilde{A}(x)$.

Theorem 1.20. *Let x be a word in a finite alphabet Σ. We have $A^-(x) \le A(x)$. If $|x| = 1$ and $|\Sigma| > 1$, then $A^-(x) = 1 < 2 = A(x)$.*

Proof. The inequality $A^-(x) \le A(x)$ follows from the inclusion $\mathrm{DFA}_\Sigma \subseteq \mathrm{partDFA}_\Sigma$. Assume $|x| = 1$ and $|\Sigma| > 1$. Without much loss of generality, $x = \mathtt{0}$ and $\Sigma = \{\mathtt{0}, \mathtt{1}, \mathtt{2}\}$. The witnessing partial DFA for $A^-(x)$ is shown in (1.2) and the witnessing DFA for $A(x, \Sigma)$ is shown in (1.3):

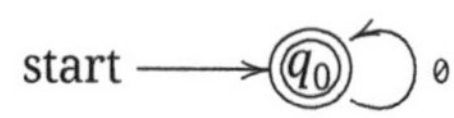

(1.2)

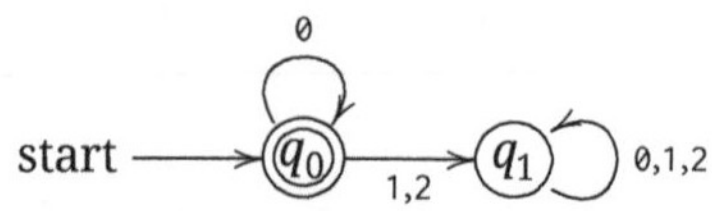

(1.3)

□

Definition 1.21 ([31, 69]). Let $L(M)$ be the language recognized by the automaton M. Let x be a finite word. The (unique-acceptance) nondeterministic automatic complexity $A_N(w) = A_{Nu}(w)$ of w is the minimum number of states of an NFA M such that M accepts w and the number of walks along which M accepts words of length $|w|$ is one.

The exact-acceptance nondeterministic automatic complexity $A_{Ne}(w)$ of w is the minimum number of states of an NFA M such that M accepts w and $L(M) \cap \Sigma^{|w|} = \{w\}$.

Frank Stephan introduced the following terminology: if $L(M) \cap \Sigma^n = \{x\}$, then we say that M accepts x *exactly*, whereas if in addition there is only one accepting walk, then we say that M accepts x *uniquely*.

Insisting that there be only one accepting walk enforces a kind of unambiguity at a fixed length. This appears to reduce the computational complexity of $A_N(x)$, compared to requiring that there be only one accepted word, since one can use matrix exponentiation. It is not known whether these are equivalent definitions (Question 1.3).

Clearly, $A_N(x) \le A(x)$. Thus, our lower bounds in, e. g., Chapter 5 for $A_N(x)$ apply to $A(x)$ as well.

Remark 1.22. We did not define "walk" above. So let us say that a sequence of states $q_1, \ldots, q_{k+1}$, $k \ge 0$, from the set of states Q in an NFA M is an *accepting state sequence* if:
- q_1 is an initial state of M,
- q_{k+1} is a final state of M, and
- each $q_i \in \delta^*(q_{i-1}, b_i)$ for some $b_i \in \Sigma$.

In this case we also say that we have an accepting state sequence for the word $b_1 \ldots b_k$. Then we require that if $x = x_1 \ldots x_n$, each $x_i \in \Sigma$, then the number of accepting state sequences for x is 1.

M accepts x uniquely if M accepts x, and there is a unique walk in M of length $|x|$ from the start state q_0 to the final state q_f.

In type theory, walks may be defined inductively as follows.
- For each state q, there is a walk called the *nil walk* from q to q.
- If there is a walk p from q_1 to q_2 and an edge e from q_2 to q_3, then there is a walk, called the walk constructed from p and e, from q_1 to q_3.

The length $|p|$ of a walk p is defined inductively as follows.
- The length of the nil walk is 0.
- The length of the walk constructed from p and e is $|p| + 1$.

We can then define the notion that M reads the word x along the walk p. See Exercise 1.19 and Exercise 2.3.

1.2.1 Formal proofs

Proof assistants such as Lean be used to formally verify mathematical theorems. Automatic complexity is no exception. Formally proving seemingly "obvious" statements can be rather instructive. Recall the word-counting nondeterministic automatic complexity $A^{Ne}(w)$, the least number of states of an NFA that accepts w and no other word of length $|w|$. This raises a couple of questions:

1. Do we know that such an NFA with a finite number of states exists?
2. What exactly should we mean by NFA?
3. How does this all fit with dependent type theory, which is used instead of set theory by proof assistants?

Traditionally, NFAs are declared to have a single start state and a set of accept states. The Lean `mathlib` library as of August 2022 defines NFA to have a set of start states and a set of accept states. This has certain advantages: the reverse of an NFA is an NFA, an NFA on the empty set of states exists, and the "folding" of the transition function δ to δ^* becomes more natural in a way. On the other hand, NFAs that are witnesses for automatic complexity may be assumed to have only one start state and one final state. Such NFAs are also closed under reversal. In Lean code we call them `NFA'`. In this book, most exercises have solutions in Lean. For such problems, the reader may try to either write their own solution in Lean or write a traditional mathematical proof by hand.

Lemma 1.23. *If $\delta_1 \vdots Q_1 \times \Sigma_1 \to Q_1$ and $\delta_2 \vdots Q_2 \times \Sigma_2 \to Q_2$ with $\delta_1 \subseteq \delta_2$, then $\delta_1^* \subseteq \delta_2^*$.*

Proof. We prove $(\forall x \in \Sigma^*)\varphi_x$, where φ_x is

$$(\forall q, r \in Q_1)(\delta_1^*(q,x) = r \to \delta_2^*(q,x) = r),$$

by structural induction on x.

As a basis step, φ_ε amounts to $q = r \to q = r$, which is true.

Now assume φ_y and let us prove $\varphi_{y::a}$. Assume $\delta_1^*(q,ya) = r$. In particular, $\delta_1^*(q,y)$ is defined and is some state s with $\delta_1(s,a) = r$. By φ_n, we have $\delta_2^*(q,y) = s$ as well. Since $\delta_1 \subseteq \delta_2$, we also have $\delta_2(s,a) = r$. Then $\delta_2^*(q,ya) = \delta_2(\delta_2^*(q,y),a) = \delta_2(s,a) = r$, as desired. □

In Lemma 1.24, we do not need to assume $Q \subseteq Q'$ or $\Sigma \subseteq \Sigma'$, but they follow from $\delta \subseteq \delta'$ with mild assumptions about all states and symbols being used nontrivially.

Lemma 1.24. *If $M = (Q,\Sigma,\delta,q_0,F)$ and $M' = (Q',\Sigma',\delta',q_0,F')$ are partial DFAs with $\delta \subseteq \delta'$ and $F \subseteq F'$, then $L(M) \subseteq L(M')$.*

Proof. By Lemma 1.23, $\delta^* \subseteq \delta'^*$. Then

$$\begin{aligned} L(M) &= \{x \mid \delta^*(q_0, x) \in F\} \\ &\subseteq \{x \mid \delta'^*(q_0, x) \in F\} \quad \text{since } \delta^* \subseteq \delta'^* \\ &\subseteq \{x \mid \delta'^*(q_0, x) \in F'\} \quad \text{since } F \subseteq F' \\ &= L(M'). \end{aligned}$$

□

Lemma 1.25. *If $\delta : Q \times \Gamma \to Q$ and δ' is the restriction of δ to $Q \times \Delta$, i. e.,*

$$\delta' = \{((q, b), r) \mid \delta(q, b) = r \text{ and } b \in \Delta\},$$

then for any $x \in (\Gamma \cap \Delta)^*$ and any $q \in Q$, $\delta^*(q, x) = \delta'^*(q, x)$.

Proof. We prove this by structural induction. If $x = \varepsilon$, then $\delta^*(q, x) = q = \delta'^*(q, x)$. Now assume $y \in (\Gamma \cap \Delta)^*$ and $q \in Q$, with $\delta^*(q, y) = \delta'^*(q, y)$. Let $a \in \Gamma \cap \Delta$. Then

$$\begin{aligned} \delta'^*(q_0, y :; a) &= \delta'(\delta'^*(q_0, y), a) \\ &= \delta'(\delta^*(q_0, y), a) \quad \text{by i. h.} \\ &= \delta(\delta^*(q_0, y), a) \quad \text{since } a \in \Gamma \cap \Delta. \end{aligned}$$

□

Remark 1.26 (For reverse mathematics aficionados). In Lemma 1.25, we have proved by induction on x the statement

$$\delta^*(q_0, x) = \delta'^*(q_0, x).$$

This is a Δ^0_1 statement about x in the sense of Simpson [72]. In reverse mathematics, researchers like to use induction on $\mathbb{N}$, however. We may also say that we have proved by induction on n the statement

$$\forall x(|x| = n \to \delta^*(q_0, x) = \delta'^*(q_0, x)).$$

This is on its face a Π^0_1 statement, but the quantifier is bounded, so it is Δ^0_1, again in the sense of Simpson [72].

Theorem 1.27. *Let Γ and Δ be alphabets. Let $x \in \Gamma^* \cap \Delta^*$. Then:*
1. $A^-(x, \Gamma) = A^-(x, \Delta)$;
2. $A_N(x, \Gamma) = A_N(x, \Delta)$.

Proof. Item 1: Let x be a word and let n be its length. By symmetry, it suffices to show that $A^-(x, \Delta) \leq A^-(x, \Gamma)$. Thus, let $M = (Q, \Gamma, \delta, q_0, F)$ be a partial DFA that uniquely accepts x,

$$L(M) \cap \Gamma^n = \{x\},$$

with $|Q| = A^-(x,\Gamma)$. Let $M' = (Q, \Gamma\cap\Delta, \delta', q_0, F)$ be the partial DFA where δ' is the restriction of δ to $Q \times \Delta$. In other words, M' is obtained from M by deleting any edge whose label is in $\Gamma \setminus \Delta$.

By Lemma 1.25, $\delta^*(q_0, x) = \delta'^*(q_0, x)$. Since $x \in L(M)$, $\delta^*(q_0, x) \in F$ and hence $\delta'^*(q_0, x) \in F$, i. e., $x \in L(M')$.

We have $L(M') \subseteq L(M)$ by Lemma 1.24. Hence,

$$\{x\} \subseteq L(M') \cap (\Gamma\cap\Delta)^n \subseteq L(M) \cap (\Gamma\cap\Delta)^n \subseteq L(M) \cap \Gamma^n = \{x\},$$

so $L(M') \cap (\Gamma\cap\Delta)^n = \{x\}$, i. e., M' uniquely accepts x. So $A^-(x,\Delta) \le |Q| = A^-(x,\Gamma)$.

Item 2: Note that if M is an NFA, then M' is an NFA and the same argument applies. □

In light of Theorem 1.27, we write $A^-(x) = A^-(x,\Sigma)$ whenever $x \in \Sigma^*$.

Theorem 1.28. *$A(x,\Sigma)$ depends on Σ. Specifically:*
1. *There exists $x \in \{0\}^*$ such that $A(x,\{0\}) \ne A(x,\{0,1\})$.*
2. *There exists $x \in \{0,1\}^* \setminus (\{0\}^* \cup \{1\}^*)$ such that $A(x,\{0\}) \ne A(x,\{0,1\})$.*

Proof. Item 1: Let $x = 0$. We have $A(0,\{0\}) = 1$, but $A(0,\{0,1\}) = 2$.

Item 2: Let $x = 01$. We have $A(x,\{0,1\}) = 2$ as shown in the following *state diagram*:

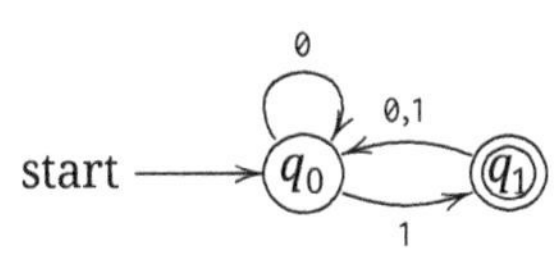

(1.4)

However, $A(x,\{0,1,2\}) = 3$, since the automaton in (1.4) is a nontotal DFA, and we cannot do better than using its dead-state completion in (1.5):

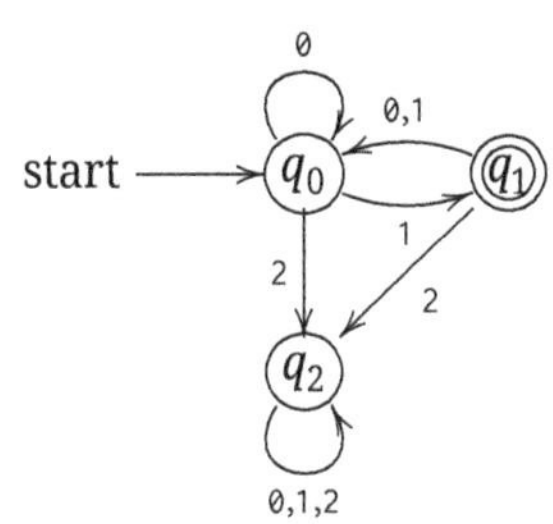

(1.5) □

Totality can always be achieved by adding at most one extra “sink” or “dead state” from which we cannot escape and cannot reach the final state.

Definition 1.29. Let $M = (Q, \Sigma, \delta, q_0, F)$ be a partial DFA. The *dead-state completion* $M_d = (Q \cup \{q_d\}, \Sigma, \delta_d, q_0, F)$ is defined as follows. The state q_d is chosen formally so that $q_d \notin Q$.

The function $\delta_d : Q \cup \{q_d\} \times \Sigma \to Q \cup \{q_d\}$ is defined by

$$\delta_d(q,b) = \begin{cases} \delta(q,b) & \text{if } \delta(q,b) \text{ is defined,} \\ q_d & \text{otherwise.} \end{cases}$$

Remark 1.30. If some convention on how states are chosen is observed, such as $Q = \{q_0, \dots, q_{|Q|-1}\}$ with $q_i = i$, then the dead-state completion is unique; otherwise it is only unique up to isomorphism.

Lemma 1.31. *Let M_d be the dead-state completion of a partial DFA M. We have $\delta_d^*(q_d, x) = q_d$ for any $x \in \Sigma^*$.*

Proof. We prove this by induction on x. By definition of δ^*, $\delta_d^*(q_d, \varepsilon) = q_d$ as desired. Now assume the result for y and let $a \in \Sigma$. Since $q_d \notin Q$, $\delta(q_d, a)$ is undefined, and hence

$$\delta_d^*(q_d, y :; a) = \delta_d^*(q_d, ya) = \delta_d(q_d, a) = q_d. \qquad \square$$

Lemma 1.32. *If M is a partial DFA and M' the dead-state completion of M, then $L(M) = L(M')$.*

Proof. Since $\delta \subseteq \delta_d$, we have $L(M) \subseteq L(M')$ by Lemma 1.24. Now suppose for contradiction that $x \in L(M') \setminus L(M)$. Since $x \in L(M')$, we have $\delta_d^*(q_0, x) \in F$, and since $x \notin L(M)$, $\delta^*(q_0, x) \notin F$. Hence, $\delta^*(q_0, x) \neq \delta_d^*(q_0, x)$.

Since certainly $\delta^*(q_0, \varepsilon) = \delta_d^*(q_0, \varepsilon)$, there must exist $w, z \in \Sigma^*$ and $a \in \Sigma$ such that $x = zaw$, and $\delta^*(q_0, z) = \delta_d^*(q_0, z)$ but $\delta^*(q_0, za) \neq \delta_d^*(q_0, za)$.

Let $r = \delta^*(q_0, z)$. Then this says that $\delta(r, a) \neq \delta_d(r, a)$. By the definition of δ_d this means that $\delta_d(r, a) = q_d$ and hence $q_d = \delta_d^*(q_0, za)$. But then by Lemma 1.31, $\delta_d^*(q_0, x) = \delta^*(q_d, w) = q_d \notin F$, which is a contradiction. $\square$

Theorem 1.33. *Let $x \in \Sigma^*$. We have:*
1. $A^-(x) \leq A(x)$,
2. $A(x) \leq A^-(x) + 1$.

Proof. Item 1 follows from the fact that each total DFA is, in particular, a partial DFA.

For Item 2, let M be a partial DFA uniquely accepting x, whose set of states Q satisfies $|Q| = A^-(x)$. Let M' be the dead-state completion of M, with set of states Q' satisfying $|Q'| = |Q| + 1$. By Lemma 1.32, $L(M') = L(M)$, so $A(x) \leq |Q'| = |Q| + 1 = A^-(x) + 1$. $\square$

Both possibilities allowed by Theorem 1.33 occur. On the one hand, $A^-(\varepsilon) = 1 = A(\varepsilon)$. On the other hand, $A^-(0) = 1 < A(0) = 2$.

Definition 1.34. For each word x, the *reverse* x^R is defined by structural induction on x:

$$\varepsilon^R = \varepsilon,$$
$$(y :; a)^R = a :: (y^R).$$

For example, $\texttt{00110}^R = \texttt{01100}$.

Definition 1.35. A function $C : \Sigma^* \to \mathbb{N}$ is *reversible* if for all $x \in \Sigma^*$, $C(x) = C(x^R)$.

Theorem 1.36. *The function $A_\Sigma(x) = A(x, \Sigma)$ is not reversible.*

Proof. For instance,

$$A(011100) = 4 < 5 = A(001110). \tag{1.6}$$

Equation (1.6) was verified by a computer program; for the idea and a partial proof see Figure 1.2. □

From Theorem 1.36 we can deduce that A^- is not reversible either.

Theorem 1.37. *We have $A^-(011100) = 3 < 4 \le A^-(001110)$.*

Proof. Since the DFA in Figure 1.2 has four states, including a dead state that is not shown, by not including that dead state we see that $A^-(011100) = 3$. On the other hand, from the mere fact that $A(001110) = 5$ we deduce by Theorem 1.33 Item 2 that $A^-(001110) \ge A(001110) - 1 = 4$. □

The *reverse* of a language L is $L^R := \{x^R \mid x \in L\}$.

Theorem 1.38. *The function A_N is reversible.*

Proof. Let M be an NFA witnessing $A_N(x)$. In particular, there is only one accepting walk $\pi = q_0 q_1 \dots q_n$ where $|x| = n$, $q_0, \dots, q_n$ are states of M, and each $q_{i+1} \in \delta(q_i)$ in M. We may assume that q_n is the only accepting state of M, so that M is of the form

$$M = (Q, \Sigma, \delta, q_0, \{q_n\}).$$

Let δ^R be defined by $q_1 \in \delta^R(q_2) \iff q_2 \in \delta(q_1)$. Let

$$M^R = (Q, \Sigma, \delta^R, q_n, \{q_0\}).$$

The only accepting walk in M^R of the same length as π is $q_n q_{n-1} \dots q_0$.

Indeed, $r_0, \dots, r_n$ is an accepting walk in M^R iff $r_0 = q_n$, $r_{i+1} \in \delta^R(r_i)$ for each i, and $r_n = q_0$; iff $r_0 = q_n$, $r_i \in \delta(r_{i+1})$ for each i, and $r_n = q_0$; and iff $r_n, \dots, r_0$ is an accepting walk in M.

Therefore, M^R witnesses that $A_N(x^R) \le A_N(x)$. By symmetry, $A_N(x^R) = A_N(x)$. □

Adding to Theorem 1.33, we have

$$A_N(x) \le A^-(x) \le A(x) \le A^-(x) + 1.$$

The minimum complexity $A_N(w) = 1$ is only achieved by words of the form a^n, where a is a single letter.

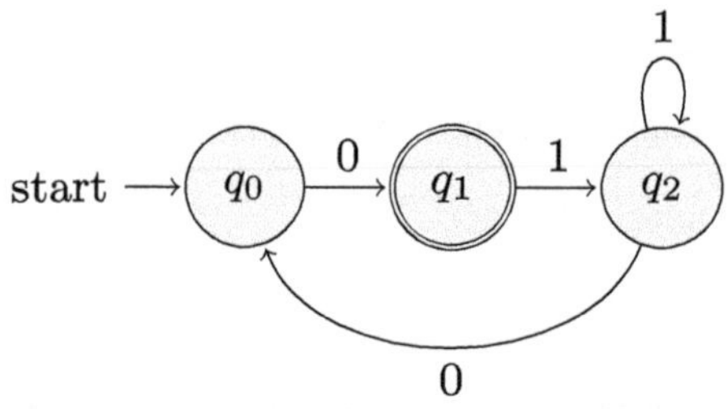

Figure 1.2: A "power-bordered" witnessing automaton for the inequality $A(011100) \leq 4$. All missing transitions go to a dead state q_3, which is not shown.

1.2.2 Subword inequalities

Shallit and Wang proved the special case $x = y$, $z = \varepsilon$ of Item 1 of Theorem 1.41 and mentioned that the simple proof thereof was shown to them by Kai Salomaa at the Descriptional Complexity of Automata, Grammars and Related Structures (DCAGRS) workshop held in London, Ontario during July 27–29, 2000.

Definition 1.39. Let $C : \Sigma^* \to \mathbb{N}$. We say that C satisfies the first subword inequality, bi-monotonicity, if for all $x, y, z \in \Sigma^*$ we have (1.7), and the second subword inequality, bi-shortness, if for all $x, y, z \in \Sigma^*$ we have (1.8):

$$C(y) \leq C(xyz), \tag{1.7}$$

$$C(xyz) \leq |x| + C(y) + |z|. \tag{1.8}$$

We also find it useful to decouple these subword inequalities as follows.

Definition 1.40. Inequality (1.9) is called *shortness* and inequality (1.10) is called *monotonicity*:

$$C(x :; a) \leq C(x) + 1, \tag{1.9}$$

$$C(x) \leq C(x :; a). \tag{1.10}$$

The constructor of Σ^* maps a and x to $x :; a$ (alternatively, $a :: x$). The constructor of the natural numbers maps n to $n + 1$. Thus, the inequality $C(x :; a) \leq C(x) + 1$ converts the constructor for words into that for natural numbers.

At the level of automata, monotonicity arises from *restrictability* and shortness from *optimal extendibility*. *Universal extendibility* refers to a situation where each witness can be extended by adding at most one state to permit one more symbol. This fails for A^-. However, *optimal extendibility* means that at least some witness can be extended, and this holds for A^-. Combining shortness and monotonicity we have $A(x :; a) \in \{A(x), A(x) + 1\}$.

Theorem 1.41. *Let Σ^* be an alphabet.*

1. *The automatic complexity function A_Σ given by $A_\Sigma(x) = A(x,\Sigma)$ satisfies the first subword inequality.*
2. *The function A^- satisfies the first subword inequality.*
3. *The function A_N (Chapter 2) satisfies the first subword inequality.*

Proof. Item 1: Let

$$M = (Q, \Sigma, \delta, q_0, F)$$

be a DFA with $A(xyz)$ many states, such that

$$L(M) \cap \Sigma^{|xyz|} = \{xyz\}. \tag{1.11}$$

Let

$$M' = (Q, \Sigma, \delta, \delta^*(q_0, x), \{\delta^*(q_0, xy)\}).$$

Then M' is also a partial DFA. It remains to show that $L(M') \cap \Sigma^{|y|} = \{y\}$. It is clear that $y \in L(M') \cap \Sigma^{|y|}$. To show $L(M') \cap \Sigma^{|y|} \subseteq \{y\}$, suppose for a contradiction that $y' \neq y$, $y' \in L(M') \cap \Sigma^{|y|}$. Then $xy'z \neq xyz$ and $xy'z \in L(M) \cap \Sigma^{|xyz|}$, contradicting (1.11).

Item 2: Note that if M is a total DFA, then so is M', since the function δ is unchanged.

Item 3: If $q_0 \dots q_n$ is the unique accepting walk, let

$$M' = (Q, \Sigma, \delta, q_{|x|}, \{q_{|xy|}\}).$$

If M' has another accepting walk $q_{|x|}p_1 \dots p_{|y|-1}q_{|xy|}$, then

$$q_0 \dots q_{|x|}p_1 \dots p_{|y|-1}q_{|xy|} \dots q_n$$

would be another accepting walk in M. In fact, we could use this proof for Item 1 and Item 2 as well. □

Theorem 1.42. *The function A_N satisfies the second subword inequality.*

Proof. Let $x, y, z \in \Sigma^*$ be given. We must show that $A_N(xyz) \leq |x| + A^-(y) + |z|$. Let $M = (Q, \Sigma, \delta, q_0, F)$ uniquely accept y, with $|Q| = A_N(y)$.

Let $M' = (Q', \Sigma, \delta', q_0', F')$, where

$$Q' = Q \cup \{r_0, \dots, r_{|x|-1}\} \cup \{s_0, \dots, s_{|z|-1}\}.$$

Here all states listed for Q' are distinct, and the unions are disjoint. Note that $|Q'| = |x| + A_N(y) + |z|$. Let

$$q_0' = \begin{cases} q_0, & \text{if } |x| = 0, \\ r_0, & \text{otherwise.} \end{cases}$$

Let

$$x = x_0 \dots x_{|x|-1},$$
$$z = z_0 \dots z_{|z|-1},$$

with each $x_i, z_i \in \Sigma$. The function δ' is equal to δ when restricted to Q. If $|x| > 0$, we let $\delta'(r_i, x_i) = \{r_{i+1}\}$ for each $0 \le i < |x| - 1$. Let $\delta'(s_i, z_{i+1}) = \{s_{i+1}\}$, $\delta'(r_{|x|-1}, x_{|x|-1}) = \{q_0\}$, and $\delta'(q_f, z_0) = \{s_0\}$ for $q_f \in F$. Let

$$F' = \begin{cases} F, & \text{if } |z| = 0, \\ \{s_{|z|-1}\}, & \text{otherwise.} \end{cases}$$

□

Definition 1.43. A word x is *periodic* if $x \neq \varepsilon$ and x is an n-power, $x = y^n$ for some integer $n > 1$.

A word that is not periodic is called *aperiodic* or *primitive*. A primitive binary word starting with `0` is called a *Lyndon word* [12, 55].

The pleasant properties of some of our main objects of study are summarized in Table 1.1.

Definition 1.39 concerns shortness: how complexity can be bounded as we construct a word. In the proof assistant Lean, the constructor is called `cons`. We also need a base case, `nil`.

Theorem 1.44 (Nillability). *For any finite alphabet* Σ, $A_\Sigma(\varepsilon) = 1$.

Proof. If $\Sigma = \{0, 1\}$, consider the following DFA:

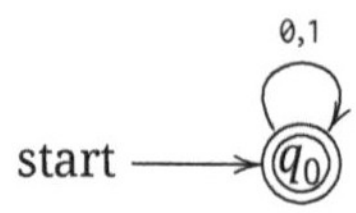

In general, let $M = (Q, \Sigma, \delta, q_0, F)$, where $Q = \{q_0\} = F$ and $\delta(q_0, a) = q_0$ for all $a \in \Sigma$. Then $L(M) \cap \Sigma^0 = \{\varepsilon\}$. □

Table 1.1: Pleasant properties of A, A^-, and A_N.

	First subword inequality	Independent of Σ	Reversible
A	Yes (Theorem 1.41)	No (Theorem 1.28)	No (Theorem 1.36)
A^-	Yes (Theorem 1.41)	Yes (Theorem 1.27)	No (Theorem 1.37)
A_N	Yes (Theorem 1.41)	Yes (Theorem 1.27)	Yes (Theorem 1.38)

Lemma 1.45. *Let $n \geq 1$, let $s = (s_0, \ldots, s_{n-1})$ be a sequence of natural numbers such that $s_0 = 1$, and for each i, $s_{i+1} \in \{s_i, s_i + 1\}$. Then $\{s_k \mid k < n\} = [1, s_{n-1}]$.*

Proof. Note that $s_i \leq s_{i+1}$ for each i. Hence, $1 \leq s_i \leq s_{n-1}$ for each $i < n$, so $\{s_k \mid k < n\} \subseteq [1, s_{n-1}]$. For the other direction, proceed by induction on n. If $n = 1$, $[1, s_{n-1}] = \{1\} = \{s_k \mid k < n\}$ as desired. Suppose $[1, s_{n-1}] \subseteq \{s_k \mid k < n\}$ and let us show $[1, s_n] \subseteq \{s_k \mid k \leq n\}$. Let $1 \leq t \leq s_n$. Either $t \leq s_{n-1}$, in which case we are done by the induction hypothesis, or $t = s_n$, in which case trivially we are done. □

Theorem 1.46. *Let Σ be a finite alphabet and let $C : \Sigma^* \to \mathbb{N}$. If C satisfies monotonicity* (1.10) *and shortness* (1.9), *then C satisfies convexity: for each k, x, y, if $C(x) < k < C(y)$, then there is some z with $C(z) = k$ (Definition 4.10).*

Proof. It suffices to prove that for each word x, $\{C(x \upharpoonright j) \mid 0 \leq j \leq |x|\} = [C(\varepsilon), C(x)]$. Then Lemma 1.45 finishes the proof. □

Theorem 1.47. *For any word x and symbol a, we have $A_N(x :; a) \in \{A_N(x), A_N(x) + 1\}$.*

Proof. By Theorem 1.46, it suffices to show that A_N satisfies monotonicity and shortness, which was shown above (Theorem 1.41 and Theorem 1.42). □

Theorem 1.48. *Let Σ be an alphabet and let $x \in \Sigma^*$. Then $A(x, \Sigma) \leq |x| + 1$.*

Proof. If $x = \varepsilon$, it follows from Theorem 1.44. Otherwise, we can use a construction with a large cycle and one dead state. □

Theorem 1.49. *Let $x \in \Sigma^*$ and let α be a nonnegative rational of the form $k + \ell/|x|$, with k, ℓ integers. We have*
1. $A(x^\alpha, \Sigma) \leq |x| + 1$;
2. $A^-(x^\alpha) \leq |x|$.

Proof. Let $x = a_1 \ldots a_n$ have length n, with each $a_i \in \Sigma$. Let M be the DFA $(Q, \Sigma, \delta, q_0, F)$, where $Q = \{q_0, \ldots, q_n\}$, all q_i are distinct, $F = \{q_\ell\}$, and δ is given by

$$\delta(q_i, a_{i+1}) = q_{i+1}, \quad 0 \leq i < n - 1,$$
$$\delta(q_{n-1}, a_n) = q_0.$$

For all other pairs $(q, b) \in Q \times \Sigma$ we let $\delta(q, b) = q_n$, so that q_n is a dead state.

Inductively, we have $\delta^*(q_0, a_1 \ldots a_j) = q_j$ for $j < n$ and $\delta^*(q_0, x) = q_\ell$, so $x \in L(M)$.

Let $|y| = n$ with $y \neq x$. Then there is some minimal i such that $y(i) \neq x(i)$. It follows that $\delta^*(q_0, y(0) \ldots y(i-2)) = q_{i-1}$ but $\delta^*(q_0, y(0) \ldots y(i-1)) = q_n$, so $\delta^*(q_0, y) = q_n$ and $y \notin L(M)$. Consequently, $L(M) \cap \Sigma^n = \{x\}$, so $A(x^\alpha, \Sigma) \leq |Q| = n + 1$. □

To prove the special case of Theorem 1.49 where $|x| = 1$ and (hence) $\alpha \in \mathbb{N}$ in great detail is Exercise 1.14.

Theorem 1.50. *Let $x, y, z \in \Sigma^*$, $x = x_1, \ldots, x_{|x|}$, $y = y_1, \ldots, y_{|y|}$, $z = z_1, \ldots, z_{|z|}$, with $x_i, y_i, z_i \in \Sigma$. Let $\alpha \in \mathbb{Q}$, $\alpha = k + \ell/|y|$, with k, ℓ nonnegative integers. Then $A^-(xy^\alpha z) \le |x| + |y| + |z|$.*

Proof. We combine the constructions in the second subword inequality for A_N in Theorem 1.42 and in Item 2 of Theorem 1.49, and have

$$A^-(xy^\alpha z) \le |x| + A^-(y^\alpha) + |z| \le |x| + |y| + |z|.$$

□

Example 1.51. Let $|x| = 1$, $|y| = 3$, $\ell = 2$, $|z| = 2$. Then the NFA from the proof of Theorem 1.50 is shown in (1.12). If $z_1 = y_3$, then (1.12) is not deterministic. However, in that case we can reduce the number of states even further, as shown in (1.13):

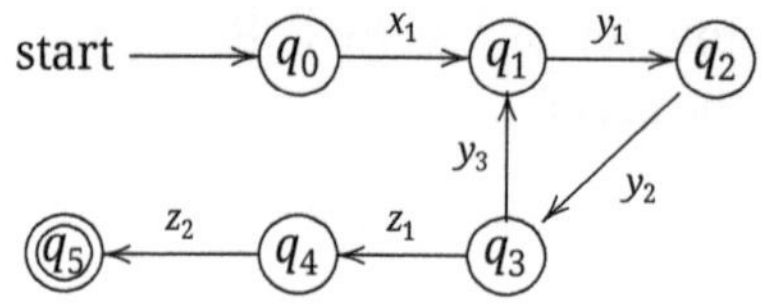

(1.12)

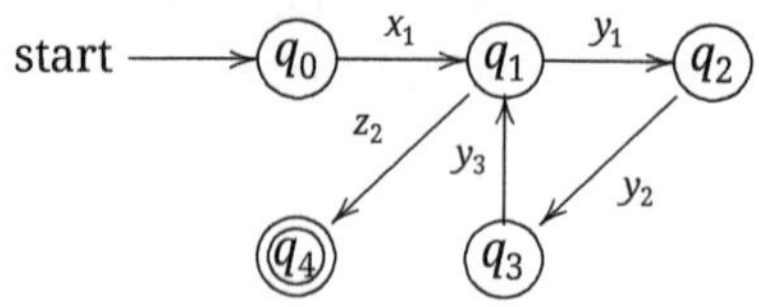

(1.13)

Similarly if the symbols $z_1, \ldots, z_t$ agree with y^α for some $t \ge 1$.

Remark 1.52. In Theorem 1.50, there is no loss of generality in assuming $z_1 \ne y_{\ell+1}$. We can refactor $xy^\alpha z = x'y'^\beta z'$ with $|x'y'z'| \le |xyz|$ to make either $z_1' \ne y_{\ell'+1}'$ or $z' = \varepsilon$. If there exists t with $z_t \ne y_{\ell+t}$, then we can choose a minimal such t and let $z' = z_t z_{t+1} \ldots z_{|z|}$. Otherwise, we can let $z' = \varepsilon$.

Corollary 1.53. *Let Σ be a finite alphabet. The function $A_\Sigma : \Sigma^* \to \mathbb{N}$ given by $A_\Sigma(x) = A(x, \Sigma)$ is computable.*

Proof. By Theorem 1.48, a brute force search among DFAs with at most $|x| + 1$ states suffices to find an automaton witnessing $A(x, \Sigma)$ and ruling out all smaller numbers of states.

□

Remark 1.54. Quantum automatic complexity, studied in [38], is also computable. This is a bit harder to prove than Corollary 1.53. However, it follows from Tarski's theorem on real closed fields.

1.3 Upper bounds

Shallit and Wang suggested two ways to save states: loops and "reuse." The idea of reuse is key to Hyde's Theorem 2.2. We now start to use these ideas.

Theorem 1.55 improves on [69, Theorem 5] by an additive term of 1.

Theorem 1.55. *Let $x \in \Sigma^*$ with $|\Sigma| = k \geq 2$ and $|x| = n$. Suppose $n > k^t + t - 1$. Then $A(x, \Sigma) \leq n + 1 - t$.*

Proof. Suppose $n > k^t + t - 1$. By Lemma 1.6, $x = a_1 a_2 \ldots a_n$ has at least $k^t + 1$ occurrences of subwords of length t. Since there are only k^t words of length k in Σ^*, some subword of length t occurs at least twice in x. Let y be a maximal-length repeated subword; it follows that $|y| \geq t$. Let u be the longest prefix of x that does not overlap with either occurrence of y. Similarly, let w be the longest suffix of x that does not overlap with either occurrence of y. Then we have

$$x = uyvw = uv'yw$$

for some v, v'. Moreover, if $w \neq \varepsilon$, then $v \neq \varepsilon$ and $v(0) \neq w(0)$. Since $yv = v'y$, by Lemma 1.13, there exist words r, s and an integer $e \geq 0$ such that

$$y = (rs)^e r, \quad v = sr, \quad v' = rs.$$

We have $|y| = e|rs| + |r|$ and

$$n = |x| = |uyvw| = |u(rs)^{e+1} rw| = |u| + (e+1)|rs| + |r| + |w|,$$

so $n - |y| = |u| + |rs| + |w|$. Then

$$A(x, \Sigma) \leq |u| + |r| + |s| + |w| + 1 = n + 1 - |y| \leq n + 1 - t$$

as witnessed by the automaton in (1.14) (where the dead state is not shown):

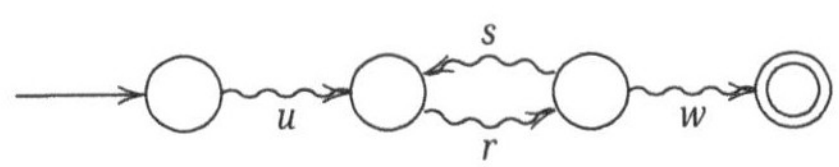

(1.14)

□

Example 1.56. The smallest nontrivial example of Theorem 1.55 is $k = 2$, $t = 1$, $n = 3$: $A(x, \Sigma) \leq 3$. If $x = \mathtt{011}$, then $u = \mathtt{0}$, $w = s = \varepsilon$, and $y = v = v' = r = \mathtt{1}$, $e = 0$. The corresponding sketch to (1.14) for $\Sigma = \{\mathtt{0}, \mathtt{1}\}$ is shown in (1.15), with full detail in (1.16):

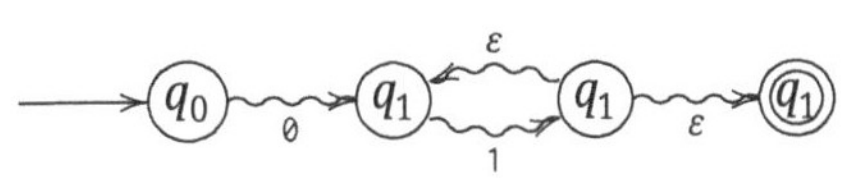

(1.15)

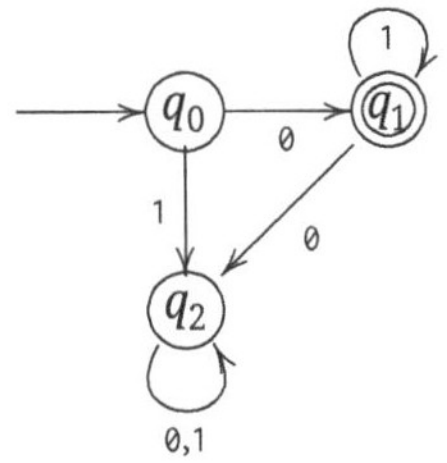

(1.16)

Theorem 1.57 (Shallit and Wang's Theorem 6). *Let* $\Sigma = \{0, 1\}$ *and* $x \in \Sigma^n$. *Then*

$$A(x, \Sigma) \leq \frac{3}{4}n + (\log n)\sqrt{\frac{n}{8}}$$

for almost all words x.

We shall not prove Theorem 1.57 here, but rather generalize it in Theorem 3.38.

1.4 Maximum-depth Dyck words

Remark 1.58. Whenever M is a DFA or an NFA, $L(M)$ is a *regular* language. The *maximum-depth Dyck words* 0^n1^n, corresponding to maximum-depth nested parentheses $(^n)^n$, are famous for providing a simple example of a nonregular language, $\{0^n1^n \mid n \geq 0\}$. It is therefore natural to investigate their irregularity in Theorem 1.60.

We define big-O notation and floor functions in Definition 1.59.

Definition 1.59. Let $f, g : \mathbb{N} \to \mathbb{N}$. We write $f(n) = O(g(n))$ if there are constants C and n_0 such that for all $n \geq n_0$, $f(n) \leq C \cdot g(n)$.

We write $f(n) = \Omega(g(n))$ if there are positive constants C and n_0 such that for all $n \geq n_0$, $C \cdot g(n) \leq f(n)$.

For $r \in \mathbb{R}$, $\lfloor r \rfloor$ is the greatest $i \in \mathbb{Z}$ with $i \leq r$ and $\lceil r \rceil$ is the least $i \in \mathbb{Z}$ with $r \leq i$.

Note that $r - 1 < \lfloor r \rfloor \leq r$ for all r.

Theorem 1.60 (Shallit and Wang's Theorem 10). *Let* $\Sigma \supseteq \{0, 1\}$ *be an alphabet. We have* $A(0^n1^n, \Sigma) = O(\sqrt{n})$.

Proof. Assume $n \geq 1$ and let $r = \lfloor \sqrt{n} \rfloor$. Then $\sqrt{n} - 1 < r \leq \sqrt{n}$, so $r^2 \leq n < (r+1)^2$. Write $n = r^2 + a$. Then $a = n - r^2 < 2r + 1$, so

$$0 \leq a \leq 2r \tag{1.17}$$

and $r \geq 1$. Noting that

$$0^n1^n = 0^a(0^r)^r1^{a+1}(1^{r+1})^{r-1},$$

we let M be the following automaton:

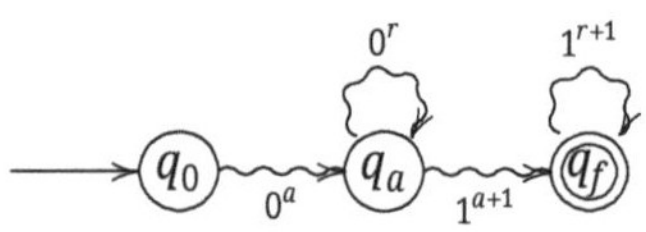

(1.18)

Here the set of states is $Q = \{q_0, \ldots, q_{|Q|-1}\}$ and $f = 2a + r$. The states occurring in the 0^r cycle are $q_a, q_{a+1}, \ldots, q_{a+r-1}$ (since $r \geq 1$). The states occurring in the 1^{a+1}

path are then $q_a, q_{a+r}, q_{a+r+1}, \ldots, q_{2a+r}$. The states occurring in the 1^{r+1} cycle are then $q_{2a+r}, q_{2a+r+1}, \ldots, q_{2a+2r}$. In addition, there is a dead state $q_{2a+2r+1}$. So M has $|Q| = 2a + 2r + 2 = 2(a + r + 1)$ states.

In a given accepting walk, if x and y are the number of times the two cycles are traversed, then the length of the walk is $2n = 2(r^2 + a) = a + rx + (a + 1) + y(r + 1)$, which reduces to

$$2r^2 = (x + y)r + (y + 1).$$

By Exercise 1.3, M uniquely accepts $0^n 1^n$.

By (1.17),

$$A(0^n 1^n) \le |Q| = 2(r + a + 1) \le 6r + 2 \le 6\sqrt{n} + 2.$$

□

Example 1.61. Shallit and Wang claimed that $2(a + r) + 1$ states are used in their construction for Theorem 1.60, so to be clear we show the case $n = 5$, where $r = 2$ and $a = 1$ in (1.19) (in this case $2(a + r) + 1 = 7$ and $2(a + r + 1) = 8$; for convenience, transitions to the dead state are indicated by a dashed line):

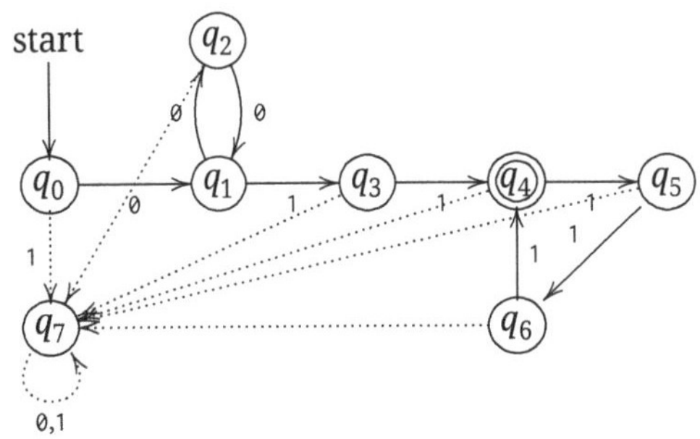

(1.19)

Of course, $n = 5$ here is too small to realize any benefits from this construction. For that we would have to go to $n = 9$, where $2(a + r + 1) = 2(0 + 3 + 1) = 8$, showing that $A(0^9 1^9, \Sigma) \le 8$.

For a DFA M with transition function δ and a word w, the processing of M on input w, i. e., the sequence of states visited by M on input w, is denoted $\delta'' w$.

Theorem 1.62. *Let $\Sigma = \{0, 1\}$. We have $A(0^n 1^n, \Sigma) \ge \sqrt{n}$ for all $n \ge 0$.*

Proof. Suppose that

$$M = (Q, \Sigma, \delta, q_0, \{q_F\})$$

is a DFA that uniquely accepts $0^n 1^n$.

Let $a \ge 0$. Suppose M has $q \le \sqrt{n} - a$ states.

We define two partial DFAs M_0 and M_1 over unary alphabets {0} and {1}, respectively:

$$M_0 = (Q_0, \{0\}, \delta_0, q_0, \{\delta(q_0, 0^n)\}),$$
$$M_1 = (Q_1, \{1\}, \delta_1, \delta(q_0, 0^n), \{q_F\}).$$

Here $Q_0 = \{\delta(q_0, 0^k) \mid 0 \le k \le n\}$ and $Q_1 = \{\delta(q_0, 0^n 1^k) \mid 0 \le k \le n\}$. Each δ_i is δ restricted to $Q_i \times \Sigma$.

Since $\sqrt{n} - \alpha < n = |0^n| = |1^n|$ (except when $n \le 1$, in which case the result is clear), some state must be repeated in $\delta_0'' 0^n$ and some state must be repeated in $\delta_1'' 1^n$. Thus, being unary and deterministic, M_0 and M_1 both consist of a path followed by a single cycle, like this:

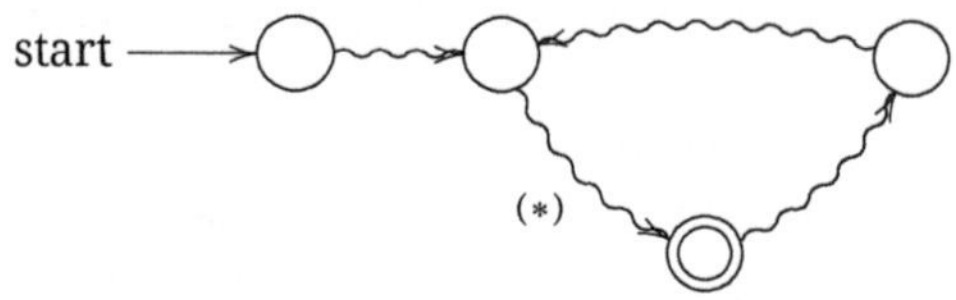

(In particular, this shows that M_0 and M_1 are total DFAs.)

Let r_i be the length of the cycle of M_i. Since some but perhaps not all states of M_i are in the cycle of M_i, we have

$$r_i \le \sqrt{n} - \alpha. \tag{1.20}$$

Let $j_i \ge 0$ be the number of times the cycle of length r_i is repeated in M_i, $i = 0, 1$. Let $t_i = n - r_i j_i$. Then t_i is the number of states of M_i appearing in the walk preceding the cycle, together with the states in (∗), so we have

$$t_i \le \sqrt{n} - \alpha. \tag{1.21}$$

Since M accepts $0^n 1^n$ uniquely and $0 \ne 1$, the equation $ar_0 + br_1 = 2n - t_0 - t_1$ must have exactly one solution $(a, b) = (j_0, j_1)$. By Exercise 1.8,

$$2n - t_0 - t_1 \le 2r_0 r_1 - r_0 - r_1. \tag{1.22}$$

We obtain a relationship between n and α as follows:

$$\begin{aligned} 2n - 2(\sqrt{n} - \alpha) &\le 2n - t_0 - t_1 \quad \text{by (1.21)} \\ &\le r_0(r_1 - 1) + r_1(r_0 - 1) \quad \text{by (1.22)} \\ &\le 2(\sqrt{n} - \alpha)(\sqrt{n} - \alpha - 1) \quad \text{by (1.20).} \end{aligned}$$

Simplifying, we have

$$n - \sqrt{n} + \alpha \le (\sqrt{n} - \alpha)(\sqrt{n} - \alpha - 1) = n - 2\alpha\sqrt{n} - \sqrt{n} + \alpha^2 + \alpha,$$

or equivalently,

$$0 \le -2a\sqrt{n} + a^2.$$

So we have shown the implication

$$A(0^n 1^n) \le \sqrt{n} - a \quad \Longrightarrow \quad 2a\sqrt{n} \le a^2.$$

For $a = 1$, this says

$$A(0^n 1^n) \le \sqrt{n} - 1 \quad \Longrightarrow \quad 2\sqrt{n} \le 1 \quad (\Longleftrightarrow n = 0).$$

For $n = 0$, $A(0^n 1^n) \ge 1$, so we conclude that for all $n \ge 0$, $A(0^n 1^n) \ge \sqrt{n}$. □

Theorem 1.62 improves on Shallit and Wang's estimate (Theorem 1.63) by an additive term of 1.

Theorem 1.63 (Shallit and Wang [69, Theorems 10 and 12]). *Let A denote deterministic automatic complexity. There is a constant n_0 such that for $n \ge n_0$,*

$$\sqrt{n} - 1 \le A(0^n 1^n) \le 6\sqrt{n} + 1.$$

We can also consider $A(0^n 1 0^n 1)$ for an example where xx is more complex than x.

Theorem 1.64 (Shallit and Wang's Theorem 13). *We have $A(0^n 1 0^n 1) = \Omega(\sqrt{n})$.*

Of course, this implies the following.

Theorem 1.65. *We have $A(0^n 1 0^n) = \Omega(\sqrt{n})$.*

Any word that has two disjoint subwords a^n, b^n for a, b single letters with $a \ne b$ must have complexity at least $\sqrt{n}$ by the same argument.

Any word that has two disjoint subwords u, v over disjoint alphabets must have complexity at least $\min\{\sqrt{|u|}, \sqrt{|v|}\}$ by the same argument.

1.4.1 Diophantine equations

Example 1.66. The importance of linear Diophantine equations to automatic complexity perhaps first really appears at length 7 for binary words. There is only one witness to the inequality $A_N(0010100) \le 3$. It is shown in (1.23):

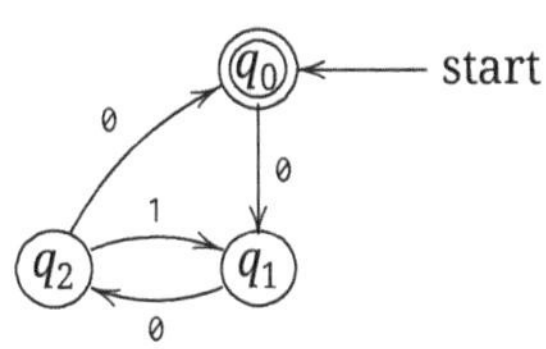

(1.23)

The reason this NFA witnesses $A_N(\texttt{0010100}) \le 3$ is the unique solvability of $2x + 3y = 7$ (with the constraint $x > 0 \implies y > 0$) over $\mathbb{N}$ (Exercise 1.5). This example also applies for the word `0101011` and generally words of the form *abcbcbd*; see Theorem 3.19.

To show that the bound $O(\sqrt{n})$ in Theorem 1.60 is tight, Shallit and Wang use Exercise 1.8. Lemma 1.67 is a generalization of Exercise 1.8.

Lemma 1.67. *Let $a_1, \dots, a_k, n$ be positive integers. Suppose that the linear Diophantine equation $a_1x_1 + \cdots + a_kx_k = n$ is solvable in nonnegative integers. If $n > (a_1 + \cdots a_k)^2 - (a_1 + \cdots + a_k)$, then the equation has at least two solutions in nonnegative integers $x_1, \dots, x_k$.*

Proof. For any $1 \le i, j \le k$, $i \ne j$, we have $a_ix_i + a_jx_j = a_i(x_i + a_j) + a_j(x_j - a_i)$. Either some pair (i, j) gives another solution in nonnegative integers, or else $x_j < a_i$ whenever $i \ne j$. Let $c_j = \min\{a_i \mid i \ne j\} - 1$. In that case, $n = a_1x_1 + \cdots + a_kx_k \le a_1c_1 + \cdots a_kc_k$. Thus, $n \le (a_1 + \cdots + a_k)^2 - (a_1 + \cdots + a_k)$. □

Theorem 1.68 is stronger than Lemma 1.67. The strategy is different: instead of using the *proof* of Exercise 1.8, we use the *statement* of Exercise 1.8.

Theorem 1.68. *Let $a_1, \dots, a_k, n$ be positive integers. Suppose that the linear Diophantine equation $a_1x_1 + \cdots + a_kx_k = n$ is solvable in nonnegative integers. If $n > \frac{1}{k-1}(a_1 + \cdots a_k)^2$, then the equation has at least two solutions in nonnegative integers $x_1, \dots, x_k$.*

Proof. We prove the contrapositive: if $a_1x_1 + \cdots + a_kx_k = n$ has exactly one solution, then $(k-1)n \le (a_1 + \cdots + a_k)^2$.

Suppose that $a_1x_1 + \cdots + a_kx_k = n$ has exactly one solution $(x_1, \dots, x_k)$. It follows that for any $1 \le i, j \le k$, $i \ne j$, the equation $a_iu + a_jv = n - \sum_{t \notin \{i,j\}} a_tx_t$ has the unique solution $(u, v) = (x_i, x_j)$.

Then Exercise 1.8 implies that for each $i \ne j$,

$$n - \sum_{t \notin \{i,j\}} a_tx_t \le 2a_ia_j - a_i - a_j.$$

Adding these $\binom{k}{2}$ many equations for $i < j$,

$$(k-1)n = \left(\binom{k}{2} - \binom{k-1}{2}\right) \cdot n \le \sum_{i<j} 2a_ia_j - a_i - a_j \le \left(\sum_i a_i\right)^2.$$

Here we have used that

$$\sum_{1 \le i < j \le k} \sum_{t \notin \{i,j\}} a_tx_t = \binom{k-1}{2} \sum_{t=1}^{k} a_tx_t,$$

which is true since for fixed t,

$$\left|\{(i,j) \mid 1 \le i < j \le k, t \notin \{i,j\}\}\right| = \binom{k-1}{2}.$$ □

Lemma 1.69 (Shallit and Wang's Lemma 14). *Let $k \ge 1$ and let $n_1, n_2, \ldots, n_k$ be positive integers, relatively prime in pairs. Let $r \ge 0$ be an integer. Define $P = \prod_{i=1}^{k} n_i$. If $r = 0$, the linear Diophantine equation*

$$\sum_{i=1}^{k} a_n \frac{P}{n_i} = \sum_{i=1}^{k} (n_i - 1)\frac{P}{n_i} - rP \tag{1.24}$$

has a unique solution $a_i = n_i - 1$, $1 \le i \le k$, in nonnegative integers. If $r \ge 1$, then (1.24) has no solutions in nonnegative integers.

Proof. We prove this by induction on k. If $k = 1$, (1.24) becomes $a_1 = n_1 - 1 - rn_1$. If $r = 0$, this equation has the unique solution $a_1 = n_1 - 1$, and if $r \ge 1$, then clearly there is no nonnegative integer solution for a_1.

Now assume the result is true for $1, 2, \ldots, k-1$. We prove it for k. Consider (1.24) mod n_k. We get

$$a_k \frac{P}{n_k} \equiv -\frac{P}{n_k} \pmod{n_k}.$$

Since the n_i are pairwise relatively prime, it follows that $a_k \equiv -1 \pmod{n_k}$. Since a_k is a nonnegative integer, we can write $a_k = jn_k - 1$ for some integer $j \ge 1$.

Now substitute $a_k = jn_k - 1$ in (1.24). After a little easy algebra, we get

$$\sum_{i=1}^{k-1} a_i \frac{P}{n_i} = \sum_{i=1}^{k-1} (n_i - 1)\frac{P}{n_i} - (j + r - 1)P. \tag{1.25}$$

By induction, since $j + r - 1 \ge 1 + r - 1 = r \ge 0$, (1.25) has a solution iff $j + r - 1 = 0$. Since $j \ge 1$ and $r \ge 0$, this can only happen if $j = 1$ and $r = 0$. So the only solution for a_k is $a_k = jn_k - 1 = n_k - 1$. Since $r = 0$, by induction there is a unique solution for a_i, $1 \le i < k$, given by $a_i = n_i - 1$. □

1.4.2 Generalized maximum-depth Dyck words

Lemma 1.70 (Shallit and Wang Lemma 15). *Let B and M be nonnegative real numbers.*
(a) *If $M^{1/k} > 2B$, then*

$$\frac{M}{M^{1/k} - B} < M^{\frac{k-1}{k}} + 2BM^{\frac{k-2}{k}}.$$

(b) *If* $0 < B < A$ *and* $k \geq 1$, *then*

$$(A - B)^k \geq A^k - kA^{k-1}B.$$

Proof. Item a: We have

$$(M^{1/k} - B)(M^{\frac{k-1}{k}} + 2BM^{\frac{k-2}{k}}) = M + BM^{\frac{k-2}{k}}(M^{1/k} - 2B) > M.$$

Item b: Apply Exercise 1.4 with $x = B/A$ and multiply by A^k. □

Let us say that a set $S \subseteq \mathbb{N}$ is *pairwise relatively prime* if for all $a, b \in S$, if $a \neq b$, then $\gcd(a, b) = 1$.

For $s \geq 1$, define $\text{findPRP}(s)$ to be the least integer n such that every interval $I \subset \mathbb{N}$ with $|I| = n$ contains a pairwise relatively prime subset of size s. Here, $|I|$ denotes the cardinality of I (which is also the length of I, plus one).

As an example, we calculate $\text{findPRP}(4)$. Part of the idea is contained in the trivial Lemma 1.71.

Lemma 1.71. *Let S be a finite subset of $\mathbb{N}$, let $p_1, p_2, \ldots$ be the sequence of all prime numbers, and let $t \in \mathbb{N}$. Suppose that for all $u, v \in S$ with $u \neq v$ and all $i < t$, $u \not\equiv v \pmod{p_i}$ and $|u - v| < p_t$. Then S is pairwise relatively prime.*

Lemma 1.72. *We have* $\text{findPRP}(4) = 6$.

Proof. The interval $[2, 6] = \{2, 3, 4, 5, 6\}$ contains no pairwise relatively prime subset of cardinality 4, so $\text{findPRP}(4) \geq 6$. For the other direction, given an interval I with $|I| = 6$, let $l < 6$ be such that for some k, $I = [6k + l, 6(k + 1) + l - 1]$. If $l \in \{0, 1, 4, 5\}$, we find using Lemma 1.71 with $p_t = 5$ that the elements of I whose remainders mod 6 are 1, 2, 3, 5 are pairwise relatively prime. If $l \in \{2, 3\}$, we find that the elements of I whose remainders mod 6 are 1, 3, 4, 5 are pairwise relatively prime. □

The following Lemma 1.73 follows easily from results of Erdős and Selfridge [19].

Lemma 1.73. *For all $\delta > 0$ and t sufficiently large we have* $\text{findPRP}(t) < t^{2+\delta}$.

Proof. Erdős and Selfridge defined $F(n, k)$ to be the largest cardinality of a pairwise relatively prime subset of the interval $[n + 1, n + k]$:

$$F(n, k) = \max\{|S| : S \subseteq [n + 1, n + k],\ S \text{ p. r. p.}\}.$$

On the one hand, $F(n, k) \leq k$. On the other hand, they proved that for all $\varepsilon > 0$, for all sufficiently large k, we have $\min_{n \geq 0} F(n, k) > k^{1/2-\varepsilon}$. Now let $k = t^{2+5\varepsilon}$ for some $\varepsilon < 1/10$. We find

$$\min_{n \geq 0} F(n, t^{2+5\varepsilon}) > (t^{2+5\varepsilon})^{1/2-\varepsilon} = t^{1+\varepsilon/2-5\varepsilon^2} = t^{1+\varepsilon(\frac{1}{2}-5\varepsilon)} > t,$$

since $\varepsilon < 1/10$. Hence, for all $0 < \varepsilon < 1/10$ and all t sufficiently large, any $t^{2+5\varepsilon}$ consecutive integers contain a small subset of cardinality $> t$. In other words, findSmall$(t) < t^{2+\delta}$, where $\delta = 5\varepsilon$. □

Theorem 1.74. *Let* $\mathtt{a}_i$, $1 \le i \le k$, *be distinct symbols. Then* $A(\mathtt{a}_1^n \mathtt{a}_2^n \dots \mathtt{a}_k^n) = O(n^{1-1/k})$, *where the constant in the big O may depend on k.*

Proof. The idea is as follows: we choose k pairwise relatively prime integers $n_i, 1 \le i \le k$, each $\le n^{1/k}$. Let $P = \prod_{i=1}^{k} n_i$. We then form a DFA similar to that in (1.18) with k cycles, one on each $\mathtt{a}_i$, $1 \le i \le k$, of size P/n_i. The case $k = 3$ is shown schematically in (1.26):

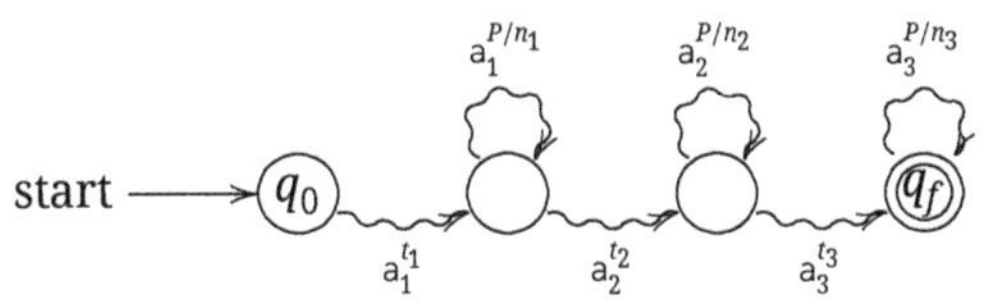

(1.26)

Each loop is preceded by a "tail" of length $t_i := n - (P/n_i)(n_i - 1) = n - P + P/n_i$. By Lemma 1.69, this DFA uniquely accepts $\mathtt{a}_1^n \mathtt{a}_2^n \dots \mathtt{a}_k^n$. The total number of states is $\le N$, where

$$N = 1 + k(n - P) + 2\sum_{i=1}^{k} \frac{P}{n_i}. \tag{1.27}$$

Since the interval

$$[\lceil n^{1/k} - k^{2+\delta} \rceil + 1, \lfloor n^{1/k} \rfloor - 1]$$

has length

$$\begin{aligned} &(\lfloor n^{1/k} \rfloor - 1) - (\lceil n^{1/k} - k^{2+\delta} \rceil + 1) + 1 \\ &= (\lfloor n^{1/k} \rfloor - 1) - (\lceil n^{1/k} - k^{2+\delta} \rceil) \\ &\ge n^{1/k} - 2 - (n^{1/k} - k^{2+\delta}) - 1 = k^{2+\delta} - 3 \ge f(k), \end{aligned}$$

by Lemma 1.73 (replacing δ by $\delta/2$ and making k larger if needed) we can choose the pairwise relatively prime numbers n_i such that $n^{1/k} - k^{2+\delta} < n_i < n^{1/k}$. Setting $A = n^{1/k}$ and $B = k^{2+\delta}$ in Item b of Lemma 1.70 we obtain

$$P = \prod_{i=1}^{k} n_i \ge n - k^{3+\delta} n^{\frac{k-1}{k}}.$$

Hence,

$$k(n - P) < k^{4+\delta} n^{\frac{k-1}{k}}. \tag{1.28}$$

On the other hand, setting $M = P$ and $B = k^{2+\delta}$ in Item a of Lemma 1.70 we obtain

$$\frac{P}{n_i} < P^{\frac{k-1}{k}} + 2k^{2+\delta} P^{\frac{k-2}{k}} \tag{1.29}$$

for all n sufficiently large. Combining equations (1.27) to (1.29), we obtain (using $P \leq n$ in the last step)

$$\begin{aligned} N &< 1 + k^{4+\delta} n^{\frac{k-1}{k}} + 2\sum_{i=1}^{k} (P^{\frac{k-1}{k}} + 2k^{2+\delta} P^{\frac{k-2}{k}}) \\ &= 1 + k^{4+\delta} n^{\frac{k-1}{k}} + 2k(P^{\frac{k-1}{k}} + 2k^{2+\delta} P^{\frac{k-2}{k}}) \\ &\leq 1 + k^{4+\delta} n^{\frac{k-1}{k}} + 2k(n^{\frac{k-1}{k}} + 2k^{2+\delta} n^{\frac{k-2}{k}}) = O(k^{4+\delta} n^{\frac{k-1}{k}}), \end{aligned}$$

as desired. □

Here $f(k, n) = O(g(k, n))$ means that there exist C, k_0, n_0 such that whenever $k \geq k_0$ and $n \geq n_0$, we have $f(k, n) \leq Cg(k, n)$.

Let us note that Theorem 1.74 is flexible: we may freely insert particular strings σ_i in between the occurrences of a_i^n as long as we add $|\sigma_i|$ many states.

We will be interested in when $O(n^{1-1/k}) \leq n$ in Theorem 1.74, that is, when

$$1 + k^{4+\delta} n^{\frac{k-1}{k}} + 2k(n^{\frac{k-1}{k}} + 2k^{2+\delta} n^{\frac{k-2}{k}}) \leq n.$$

However, this analysis must also take into account how findPRP gets small depending on δ. Erdős and Selfridge [19] used an “unpublished result of Rosser” which is supposedly in “the forthcoming book on sieve methods by Halberstam and Richert,” presumably [27]. As pointed out in [54], we can trace the result to a paper by Iwaniec [32].

1.5 Open problems

Question 1.1. By Theorem 1.47, if $k < A_N(y)$, then there is a prefix z of y with $A_N(z) = k$. Can this k can be found efficiently?

Question 1.2. In trying to prove a second subword inequality for $A = A_\Sigma$ we must consider that a dead state might be added. We have $A(xyz) \leq |x| + A(y) + |z| + 1$, using the construction from Theorem 3.44. Can we remove the +1? To find a counterexample, start with a y which requires no dead state for $A(y)$.

Question 1.3. Is $A_{Nu} = A_{Ne}$?

Question 1.4 (Due to an anonymous referee of [31]). Can the $(A(x))$ complexity of a string and its reverse be arbitrarily far apart? Formally, is it the case that

$$\forall n \exists x |A(x) - A(x^R)| \geq n?$$

Question 1.5. In a footnote, Shallit and Wang suggested an alternative notion of "automatic complexity" as well: *the minimum number of states in a DFA such that x is the lexicographically least word of length $|x|$ that is accepted.* This notion has not yet been studied in the literature.

1. What can you prove about it?
2. Does this complexity notion satisfy any of the properties in Table 1.1?

1.6 Exercises

Exercise 1.1. Find a function that satisfies the first subword inequality but not convexity.

Exercise 1.2. Find a function C that satisfies the second subword inequality but not convexity. Hint and follow-up: does the value of $C(\varepsilon)$ affect this?

Exercise 1.3. Let $r \geq 1$ be an integer. Prove that

$$\{(x,y) \in \mathbb{N} \times \mathbb{N} \mid 2r^2 = (x+y)r + (y+1)\} = \{(r, r-1)\}.$$

A complete solution to Exercise 1.3 is available at the following URL.

https://github.com/bjoernkjoshanssen/ac-exercises/

Exercise 1.4. Prove by induction on $k \in \mathbb{N}$ that if $x \in \mathbb{R}$, $0 \leq x \leq 1$, then $(1-x)^k \geq 1 - kx$.

A complete solution to Exercise 1.4 is available at the following URL.

https://github.com/bjoernkjoshanssen/ac-exercises/

(Shallit and Wang stated that $k \geq 1$ and $0 < x < 1$ were assumptions here, but working it out in Lean shows that this slightly stronger form also holds.)

Exercise 1.5. Show that the unique solution to $2x + 3y = 7$ over $\mathbb{N}$ is $(x,y) = (2,1)$.

A complete solution to Exercise 1.5 is available at the following URL.

https://github.com/bjoernkjoshanssen/ac-exercises/

Exercise 1.6. For which $n \in \mathbb{N}$ does the NFA in (1.23) uniquely accept a word of length n?

Exercise 1.7. Let $x_0, y_0, a, b \in \mathbb{Z}$ with $x_0 < b$ and $y_0 < a$. Assume that a and b are positive and relatively prime. Then the equation

$$ax + by = ax_0 + by_0 \tag{1.30}$$

has a unique solution (x,y) in nonnegative integers.

Exercise 1.7 is a complementary result to Exercise 1.8, and was used in [37]. A complete solution to Exercise 1.7, provided by Rukiyah Walker, is available at the following URL.

https://github.com/bjoernkjoshanssen/ac-exercises/

Exercise 1.8. Let $a, b, n \in \mathbb{Z}$ with $a, b \geq 1$. Suppose that the linear Diophantine equation $ax + by = n$ is solvable in nonnegative integers. If $n > 2ab - a - b$, then the linear Diophantine equation $ax + by = n$ has at least two solutions in nonnegative integers x, y.

A complete solution to Exercise 1.8, provided by Aleksander Fedorynski, is available at the following URL.

https://github.com/bjoernkjoshanssen/ac-exercises/

Exercise 1.9. For an NFA $M = (Q, \Sigma, q_0, \delta, F)$ and a symbol $a \in \Sigma$, the *head transformation* of M and a is the NFA $\mathrm{hd}(M, a) = (Q \cup \{q\}, \Sigma, q, \delta', F)$, where $q \notin Q$ and $\delta' = \delta \cup \{((q, a), q_0)\}$.
(a) Prove that $L(\mathrm{hd}(M, a)) = \{ay \mid y \in L(M)\}$.
(b) Prove that if M exactly accepts x, then $\mathrm{hd}(M, a)$ exactly accepts $a :: x$.
(c) For $a \in \Sigma$ and $x \in \Sigma^*$, use Item b to prove that $A_N^{\mathrm{word}}(ax) \leq A_N^{\mathrm{word}}(x) + 1$.
(d) Prove the subword inequality for A_N^{word},

$$A_N^{\mathrm{word}}(xy) \leq |x| + A_N^{\mathrm{word}}(y).$$

(e) Prove that $A_N^{\mathrm{word}}(\varepsilon) = 1$.
(f) Conclude that $A_N^{\mathrm{word}}(x) \leq |x| + 1$.

A complete solution to Exercise 1.9 is available at the following URL.

https://github.com/bjoernkjoshanssen/ac-exercises/

Exercise 1.10. The path-based automatic complexity $A_N(x)$ for $x \in [b]^n$ can be characterized as follows. Let us say that (h, x) covers (g, y) if $g(0) = h(0)$ and $g(n) = h(n)$, and also for all k there is an l with $(g(k), g(k+1), y(k)) = (h(l), h(l+1), x(l))$. (The idea is that the NFA generated by the state sequence h and the word x contains as a substructure the NFA generated by g and y.)

We have $A_N(x) \leq q$ iff there exists a sequence $h \in [q]^{n+1}$ such that for all $g \in [q]^{n+1}$ and $y \in [b]^n$, if $h(x)$ covers (g, y), then $g = h$ and $x = y$.

Do not prove this; instead assume it and use it to verify the facts in Table 1.2.

A complete solution to Exercise 1.10 is available at the following URL.

https://github.com/bjoernkjoshanssen/ac-exercises/

Exercise 1.11. Prove that for all $n \in \mathbb{N}$, there exists a word $x \in \{0, 1\}^n$ such that x is primitive.

Table 1.2: Some elementary calculated facts about nondeterministic, path-based automatic complexity A_N. The asterisks (*) indicate power-bordered words, for which utilizing the form *abccda* gives an optimal witness.

x	$A_N(x)$
ε	1
0	1

x	$A_N(x)$
00	1
01	2

x	$A_N(x)$
000	1
001	2
010	2
011	2

x	$A_N(x)$
0000	1
0001	2
0010	3
0011	3
0100	3
0101	2
0110	3
0111	2

x	$A_N(x)$
00000	1
00001	2
00010	3
00011	3
00100	3
00101	3
00110	3
00111	3
01000	3
01001	3
01010	2
01011	3
01100	3
01101	3
01110	3
01111	2

x	$A_N(x)$
000000	1
000001	2
000010	3
000011	3
000100	4
000101	4
000110	4
000111	4
001000	4
001001	3
001010	3
001011	4
001100	3*
001101	4
001110	3*
001111	3
010000	3
010001	4
010010	3
010011	4
010100	3
010101	2
010110	4
010111	4
011000	4
011001	4
011010	4
011011	3
011100	3*
011101	4
011110	3
011111	2

A complete solution to Exercise 1.11 is available at the following URL.

https://github.com/bjoernkjoshanssen/ac-exercises/

Exercise 1.12. A word x is *bordered* if there are no words v and $u \neq \varepsilon$ such that $x = uvu$. Prove that a nonprimitive word is bordered. Prove that for all $n \in \mathbb{N}$, there exists a word $x \in \{0,1\}^n$ such that x is not bordered.

Exercise 1.12 can be proved by extending the proof of Exercise 1.11.

Exercise 1.13. Prove that for all $n \in \mathbb{N}$, there exists a word $x \in \{0,1\}^n$ such that no prefix of x is bordered.

Exercise 1.13 can be proved by extending the proof of Exercise 1.12.

Exercise 1.14. Prove that $A(z^n, \Sigma) \leq 2$ for all $n \in \mathbb{N}$ and Σ with $z \in \Sigma$.

A complete solution to Exercise 1.14 is available at the following URL.

https://github.com/bjoernkjoshanssen/ac-exercises/

Exercise 1.15. For $k \in \mathbb{N}$, the *k-hydewalk* (named after Kayleigh Hyde) is the walk witnessing that a Kayleigh graph NFA (Theorem 2.2) uniquely accepts a given word of odd length $n = 2k + 1$. It is a function from a set of cardinality $n + 1$ to a set of cardinality $q = k + 1$: $h : [2k + 2] \to [k + 1]$, where $[a] = \{0, \ldots, a - 1\}$. Define the function h arithmetically, by cases:

$$h(i) = \begin{cases} ? & i < ? \\ ? & i \geq ? \end{cases}$$

Exercise 1.16. Let the *k-sidewalk* be given by $g : [k + 1] \to [k + 1]$, $g(i) = i$ for all i. Show that the k-hydewalk covers (in the sense of Exercise 1.10) the k-sidewalk but not vice versa.

Exercise 1.17. Show that the covering relation (Exercise 1.10) is reflexive and transitive.

Exercise 1.18. A walk g locally preserves parity at i if $g(i) \equiv i \pmod 2$. If g locally preserves parity at each i for which $g(i)$ is defined, then we say that g preserves parity. Show that the k-sidewalk (Exercise 1.16) preserves parity but the k-hydewalk (Exercise 1.15) does not. Specifically, at any $i \geq k + 1$, the k-hydewalk fails to preserve parity, but at any $i < k + 1$, it does preserve parity.

Exercise 1.19. Given a sequence (of "states") s_0, s_1, s_2, s_3, construct an unlabeled digraph having s_0, s_1, s_2, s_3 as its vertices such that (s_0, s_1, s_2, s_3) is a walk in this digraph.

A formal solution to Exercise 1.19 is available at the following URL.

https://github.com/bjoernkjoshanssen/ac-exercises/

2 Nondeterminism and overlap-free words

In this chapter we develop some properties of nondeterministic automatic complexity. As a corollary we get a strengthening of a result of Shallit and Wang [69] on the complexity of the infinite Thue–Morse word **t**. Moreover, viewed through an NFA lens we can, in a sense, characterize the complexity of **t** exactly. A main technical idea is to extend the following result, which says not only that squares, cubes, and higher powers of a word have low complexity, but also that a word completely free of such powers must conversely have high complexity.

Theorem 2.1 (Shallit and Wang [69, Theorem 9]). *Suppose $w \in \Sigma^*$ is kth-power-free for some integer $k \geq 2$, i. e., w contains no subword of the form x^k with $x \neq \varepsilon$. Then $A(w) \geq \frac{|w|+1}{k}$.*

The way we strengthen their results is by considering a variation on square-freeness and cube-freeness, *overlap-freeness*. This notion also goes by the names of *irreducibility* and *strong cube-freeness* in the combinatorial literature. We also take an idea from [69, Theorem 8] and use it to show that the natural decision problem associated with nondeterministic automatic complexity is in the complexity class $\mathrm{E} = \mathrm{DTIME}(2^{O(n)})$. This result is a theoretical complement to the practical fact that the nondeterministic automatic complexity can be computed reasonably quickly, a fact that enabled the creation of the lookup service in [45].

2.1 Introduction

Theorem 2.2 (Hyde [30]). *The nondeterministic automatic complexity $A_N(x)$ of a string x of length n satisfies*

$$A_N(x) \leq b(n) := \lfloor n/2 \rfloor + 1.$$

Proof. Let $n = 2m + 1$ be odd and let K_m be a Kayleigh graph (Figure 2.1). For a proof by induction it is convenient to label the states in the opposite order of the usual:

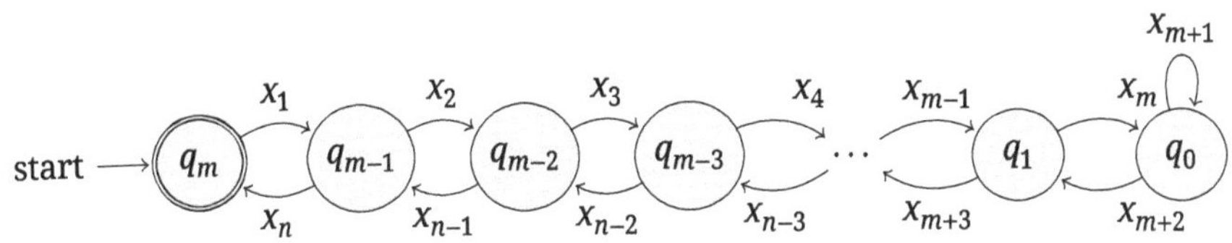

Then we just observe that there are no walks of odd length $< n$ accepted, and there is exactly one walk of odd length n accepted, and these two observations persist inductively.

For a number perspective, let $W_{m,k}$ be the number of accepting walks in K_M of length k. We show by induction on m that $W_{m,2m+1} = 1$ and $W_{m,2s+1} = 0$ whenever $s < m$.

https://doi.org/10.1515/9783110774870-002

If $m = 0$, the NFA K_m is simply

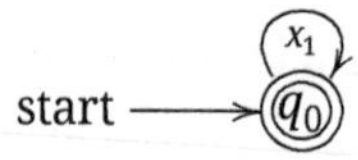

which accepts one word of length 1 and, of course, no word of shorter odd lengths.

For the inductive step, note that K_m is schematically

An accepting walk of length $2m+1$ consists of the edge labeled x_1, followed by a walk of length $2m-1$, followed by the edge labeled x_n. The walk of length $2m-1$ can stay within K_{m-1}, in which case by induction there is exactly one possible walk. Or it can venture out of K_{m-1} to visit q_0 a certain number $0 < t \le m-1$ of times; each such time it contributes a walk of length 2. Between these times, an accepting walk in K_{m-1} must be performed.

By the induction hypothesis, these walks have lengths $\ell_0, \dots, \ell_t$, where if $t > 0$ and $W_{m-1,\ell_i} > 0$, then ℓ_i is even. So if $t > 0$, they have lengths that must add up to an odd number

$$\sum_{i=0}^{t} \ell_i = 2m - 1 - 2t,$$

which is impossible. The number of walks is

$$\begin{aligned} W_{m,2m+1} &= \sum_{t=0}^{m-1} \sum_{\substack{\{\ell_0,\dots,\ell_t \ge 0: \\ \sum \ell_i = 2m-1-2t\}}} \prod_{i=0}^{t} W_{m-1,\ell_i} \\ &= \sum_{\substack{\{\ell_0 \ge 0: \\ \ell_0 = 2m-1\}}} W_{m-1,\ell_0} \\ &= W_{m-1,2m-1} \\ &= 1 \quad \text{(ind. hyp.)}. \end{aligned}$$

If $n > 0$ is even, then $x = ya$, where $|y|$ is odd; by the second subword inequality in Theorem 1.41 Item 1.41, $A_N(x) \le A_N(y) + 1$, as desired. Finally, if $n = 0$, then the result follows from $A_N(\varepsilon) = 1$. □

Definition 2.3. The *complexity deficiency* of a word x of length n is

$$D_n(x) = D(x) = b(n) - A_N(x).$$

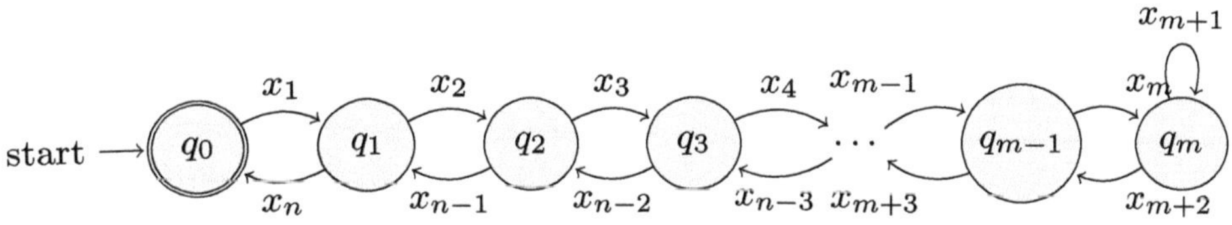

Figure 2.1: A nondeterministic finite automaton, a *Kayleigh graph*, that only accepts one string $x = x_1x_2x_3x_4 \cdots x_n$ of length $n = 2m + 1$.

Table 2.1: The number of binary words of length $0 \le n \le 23$ having nondeterministic automatic complexity k.

$n\backslash k$	1	2	3	4	5	6	7	8	9	10	11	12
23	2	6	20	58	164	430	2540	14252	80962	442278	2160662	5687234
22	2	6	20	58	164	502	2846	16024	94732	451368	2089418	1539164
21	2	6	20	58	176	496	3168	18720	108042	504794	1461670	
20	2	6	20	58	164	430	3814	23328	115896	529148	375710	
19	2	6	20	58	164	582	4996	26542	140668	351250		
18	2	6	20	58	188	598	5692	29990	136024	89566		
17	2	6	20	58	200	514	7102	37042	86128			
16	2	6	20	58	164	752	7738	34320	22476			
15	2	6	20	58	226	908	8530	23018				
14	2	6	20	58	244	1270	9668	5116				
13	2	6	20	64	250	2076	5774					
12	2	6	20	58	282	2090	1638					
11	2	6	20	58	564	1398						
10	2	6	20	64	588	344						
9	2	6	20	78	406							
8	2	6	20	130	98							
7	2	6	22	98								
6	2	6	26	30								
5	2	6	24									
4	2	6	8									
3	2	6										
2	2	2										
1	2											
0	1											

The distribution of $A_N(w)$ for w of length $n \le 23$ is given in Table 2.1. The notion of deficiency is motivated by the experimental observation that about half of all strings have deficiency 0.

2.2 Powers and complexity

In this section we shall exhibit infinite words all of whose prefixes have complexity deficiency bounded by 1. We say that such a word has a hereditary deficiency bound of 1.

2.2.1 Square-free words

Definition 2.4. A word x is a *factor* in a word y if $y = uxv$ for some words u and v. In this case we also say that y *contains* x.

We will use the following simple strengthening from DFAs to NFAs of a fact used in the proof of Theorem 2.1.

Theorem 2.5. *Let $k \in \mathbb{N}$. If an NFA M uniquely accepts w of length n and visits a state p at least $k + 1$ times during its computation on input w, then w contains a kth power.*

Proof. If $k \leq 1$ the statement is trivial, so let us assume $k \geq 2$.

Let $w = w_0 w_1 \cdots w_k w_{k+1}$, where:

- w_0 is the portion of w read before the first visit to the state p,
- w_i is the portion of w read between visits number i and $i+1$ to the state p for $1 \leq i \leq k$, and
- w_{k+1} is the portion of w read after the last visit to the state p.

Thus, $|w_i| \geq 1$ for each $1 \leq i \leq k$. It is possible to have $|w_0| = 0$ ($|w_{k+1}| = 0$) since the initial (final) state of M's computation on input w may be p.

For any permutation π on $1, \ldots, k$, M accepts $w_0 w_{\pi(1)} \cdots w_{\pi(k)} w_{k+1}$. Let $1 \leq j \leq k$ be such that w_j has minimal length and let

$$\hat{w}_j = w_1 \cdots w_{j-1} w_{j+1} \cdots w_k.$$

Then M also accepts

$$w_0 w_j \hat{w}_j w_{k+1} \quad \text{and} \quad w_0 \hat{w}_j w_j w_{k+1}.$$

By uniqueness,

$$w_0 w_j \hat{w}_j w_{k+1} = w = w_0 \hat{w}_j w_j w_{k+1},$$

so

$$w_j \hat{w}_j = \hat{w}_j w_j.$$

By Theorem 1.14, w_j and $\hat{w}_j$ are both powers of a string z. Since $|\hat{w}_j| \geq (k-1)|w_j|$, $w_j\hat{w}_j$ is at least a kth power of z, so w contains a kth power of z. □

Theorem 2.6. *Let $k \in \mathbb{N}$, $k \geq 1$. If a word w is kth-power-free, then $A_N(w) \geq \frac{|w|+1}{k}$.*

Proof. Let k and w be given. Let $q = A_N(w)$ and let M witness that $A_N(w) \leq q$. For a contrapositive proof, assume $q < \frac{|w|+1}{k}$. Thus $k < \frac{|w|+1}{q}$.

Let p be a most-visited state in M during its computation on input w. Then p is visited at least $(|w|+1)/q > k$ times, and hence at least $k+1$ times. By Theorem 2.5, w contains a kth power. □

Theorem 2.7 (Extended pigeonhole principle). *If $aq + d$ pigeons are placed in q pigeonholes, where $d > 0$, then it cannot be the case that all pigeonholes have at most a pigeons; in fact, either*
- *there is a pigeonhole with at least $a + d$ pigeons; or*
- *there is a pigeonhole with at least $a + d - 1$ pigeons and another with $a + 1$ pigeons; or*
- *there is a pigeonhole with at least $a + d - 2$ pigeons and another with $a + 2$ pigeons; or*
- *there is a pigeonhole with at least $a + d - 2$ pigeons and two others with $a + 1$ pigeons; or*
- *all pigeonholes have at most $a + d - 3$ pigeons (which is impossible if $a + d - 3 \leq a$ and $d > 0$).*

Formally, let $p, q \geq 0$ be nonnegative integers and let $d > 0$. Let $\mathcal{H}$ be a family of q disjoint subsets of $[p]$, $|\mathcal{H}| = q$, $\mathcal{H} \subseteq \mathcal{P}([p])$. Suppose that $p = aq + d$. Then it cannot be the case that $|H| \leq a$ for each $H \in \mathcal{H}$. At least one of the following must hold:
- *There is an $H \in \mathcal{H}$ with $|H| \geq a + d$.*
- *There is an $H_1 \in \mathcal{H}$ with $|H_1| \geq a + d - 1$ and an $H_2 \in \mathcal{H}$, $H_2 \neq H_1$, with $|H_2| \geq a + 1$.*
- *There is an $H_1 \in \mathcal{H}$ with $|H_1| \geq a + d - 2$ and an $H_2 \in \mathcal{H}$, $H_2 \neq H_1$, with $|H_2| \geq a + 2$.*
- *There is an $H_1 \in \mathcal{H}$ with $|H_1| \geq a + d - 2$ and $H_2, H_3 \in \mathcal{H}$, H_1, H_2, H_3 all distinct, with $|H_2|, |H_3| \geq a + 2$.*
- *Each $|H| \leq a + d - 3$.*

The last case of Theorem 2.7 is impossible if $a + d - 3 \leq a$ and $d > 0$.

Proof. Consider the maximum number of pigeons in a pigeonhole m. If $m \geq a + d$, we are in Case 1. If $m = a + d - 1$, we consider all the other pigeons and pigeonholes; there are then $q - 1$ pigeonholes and $aq + d - (a + d - 1) = a(q - 1) + 1$ pigeons. By the plain pigeonhole principle, there is a pigeonhole with at least $a + 1$ pigeons. If $m = a + d - 2$, we repeat the argument, considering the maximum number of pigeons in a pigeonhole other than a given one with the maximum number of pigeons. □

We next strengthen a particular case of Theorem 2.1 to NFAs.

Theorem 2.8. *A square-free word has deficiency 0.*

Proof. Suppose w is a word of length $n = 2k$ or $n = 2k+1$ of deficiency d. Then there is a witnessing automaton M with $q = k+1-d$ states. Since $n+1 \geq 2k+1 = 2(k+1-d)+2d-1 = 2q+(2d-1)$, by the extended pigeonhole principle (Theorem 2.7), there is a state p which is visited $2+(2d-1) = 3$ times $t_1 < t_2 < t_3$ (such times were introduced in Remark 2.18), during the $n+1$ times of the computation of M on input w (and is not visited at any other times in the interval $[t_1, t_3]$). By Theorem 2.5, w contains a square. □

Theorem 2.9 (Thue [79]). *Let Σ be an alphabet of cardinality (at least) three. There exists an infinite square-free word over Σ.*

For a contrast to Theorem 2.9, see Exercise 2.2.

Corollary 2.10. *There exists an infinite word over a finite alphabet of hereditary deficiency 0.*

Proof. There is an infinite square-free word over the alphabet $\{0, 1, 2\}$ by Theorem 2.9. Now, the result follows from Theorem 2.8. □

2.2.2 Morphisms

Definition 2.11. Let Σ be a finite alphabet. A function $\pi : \Sigma^* \to \Sigma^*$ is a *morphism* (sometimes, *homomorphism*) if it is a homomorphism with respect to concatenation, in the sense that for all x, y,

$$\pi(xy) = \pi(x)\pi(y).$$

Lemma 2.12. *Let x be a rainbow word (Definition 1.5). Then $A^-(x) = \lfloor \frac{|x|}{2} \rfloor + 1$.*

Proof. Since the Kayleigh graph (Figure 2.1) of x is deterministic, $A^-(x) \leq \lfloor \frac{|x|}{2} \rfloor + 1$. Since x is square-free, by Theorem 2.8 $\lfloor \frac{|x|}{2} \rfloor + 1 \leq A_N(x) \leq A^-(x)$. □

Definition 2.13. A finite or infinite sequence $s_0, s_1, \ldots$ from $\mathbb{N}$ is *maxshort* if for each i, $s_i \leq \max\{s_0, \ldots, s_{i-1}\} + 1$.

Theorem 2.14. *There exists a word $x \in \{0, 1\}^7$ with $A^-(x) > \frac{|x|}{2} + 1$. There exists a word $x \in \{0, 1, 2\}^5$ with $A^-(x) > \frac{|x|}{2} + 1$.*

Proof. Using computer search (see Figure 2.3, Remark 2.15), we find that the shortest binary example (starting with 0) is $x = \texttt{0001101}$, which satisfies $A^-(x) = 5 > \frac{|x|}{2} + 1$. For a ternary alphabet, `01021` is the shortest example. □

```
def maxshort(self, seq):
    currmax = -1
    for j in range(0, len(seq)):
        if seq[j] > currmax + 1: return False
        if seq[j] == currmax + 1: currmax += 1
        if currmax == self.q - 1: return True
    return True
```

Figure 2.2: Python code for maxshort sequences.

Remark 2.15. We do not know a short proof that no partial DFA with four states can uniquely accept 0001101 (Question 2.1). To avoid searching several isomorphic copies of a possible sequence of states we use maxshort sequences (Definition 2.13) as indicated in Figure 2.2.

Proposition 2.16 and Proposition 2.17 constitute another case where deterministic automatic complexity lacks a simple property that nondeterministic automatic complexity has. A morphism π is *length preserving* if it is induced by a map induced by $\pi : \Sigma \to \Sigma$ (this could be viewed as either a definition or a theorem).

Proposition 2.16. *If a morphism π is length preserving, then we have $A_N(\pi x) \leq A_N(x)$ for all x.*

Proof. Given an NFA M uniquely accepting x, we replace all edge labels ℓ by $\pi(\ell)$ to obtain an NFA M' uniquely accepting $\pi(x)$ with the same number of states as M. □

Proposition 2.17 shows that Proposition 2.16 does not hold with A_N replaced by A^-.

Proposition 2.17. *There exist a length-preserving morphism π and a word x with $A^-(\pi(x)) > A^-(x)$.*

Proof. By existential elimination from Theorem 2.14, fix a word y with $A^-(y) > n/2 + 1$, where $n = |y|$. Let $\Delta = \{\mathtt{a}_0, \mathtt{a}_1, \mathtt{a}_2, \dots \mathtt{a}_{n-1}\}$ be an alphabet of size n and let $x = \mathtt{a}_0\mathtt{a}_1\mathtt{a}_2 \dots \mathtt{a}_{n-1}$ (a rainbow word, Definition 1.5). Let $\pi : \Delta \to \Sigma$ be the morphism defined by $\pi(\mathtt{a}_i) = y_i$, where $y = y_0y_1 \dots y_{n-1}$. Then

$$\begin{aligned} A^-(\pi x) &= A^-(y) && \text{since } \pi(x) = y \\ &> \lfloor n/2 \rfloor + 1 && \text{property of } y \\ &= A^-(x) && \text{Lemma 2.12.} \end{aligned}$$

□

```
def complexityBounds(self, seq, tt, xseq, checkDeterminism, checkUnambiguity):
    if (self.isItTimeToReturn) or not self.maxshort(seq): return
    transitions = tt.copy()
    if len(xseq) == self.n:
        t = set()
        for nn in range(0, len(seq) - 1):
            p, pp = seq[nn], seq[nn + 1]
            t.add((p, pp, xseq[nn]))
            transitions[p, pp] = sum(
                [1 for bb in range(0, self.b) if (p, pp, bb) in t])
        if checkDeterminism and not self.deterministic(t): return
        if (len(seq)==self.n+1
            and checkUnambiguity and not self.unambiguous(t, seq[self.n])
        ): return
    j = len(seq) - 1
    if j > 0:
        if (transitions[seq[j - 1], seq[j]] == 0):
            transitions[seq[j - 1], seq[j]] = 1
        numPaths = numpy.linalg.matrix_power(
            transitions, j
        )[0, seq[j]]
        if (numPaths == 0) or (numPaths > 1): return
        if j == len(xseq):
            self.edges[len(t)] += 1
            if len(t) <= self.edgeLimit: # for edge complexity E_N
                print("\t\t" + str(seq) + ",")
                print("\t",max(seq)+1,"states,",len(t),"edges")
            if (self.returnWhenFound):
                self.isItTimeToReturn = True
                return
    if j < self.n:
        for p in range(self.q - 1, -1, -1):
            self.complexityBounds(
                seq + (p,), transitions, xseq,
                checkDeterminism, checkUnambiguity,
            )
        return
```

Figure 2.3: The main chunk of code we use for computing automatic complexity facts.

Remark 2.18. In Figure 2.3, `j` is the maximum input to `seq` and also the length of any relevant `xseq` going through the states in `seq`. Note that `(seq[j - 1], seq[j])` is the last transition in `seq` when `j == len(xseq) - 1`. If `numPaths == 0` or `numPaths > 1`, then we do not have the desired unique path at this prefix, so we return. The function `complexityBounds` is called from the function `complexity`:

```
def complexity(self, xseq, checkDeterminism=False, checkUnambiguity=
    False):
    tt = numpy.zeros(shape=(self.q, self.q), dtype=numpy.uint32)
    self.complexityBounds(
        (), tt, xseq, checkDeterminism, checkUnambiguity=
    checkUnambiguity
    )
```

Theorem 2.19. *We have* $A_N \neq A^-$.

Proof. Let $x = 001110$. Using computer search, we find that x is the only binary word of length at most 6 starting with 0 which has $A_N(x) \neq A^-(x)$. In fact, $A_N(x) = 3 < 4 = A^-(x)$. The inequality $A_N(x) \leq 3$ is witnessed by the state sequence 0122201. The inequality $A^-(x) \leq 4$ is witnessed by the state sequence 0122223. □

2.2.3 Overlap-free words

Definition 2.20 (Thue–Morse morphism). Let $\Sigma = \{0, 1\}$. The *Thue–Morse morphism* is the unique morphism $\mu : \Sigma^* \to \Sigma^*$ satisfying

$$\mu(0) = 01, \quad \mu(1) = 10.$$

Definition 2.21. The infinite Thue–Morse word

$$\mathbf{t} = t_0 t_1 \cdots = 0110\,1001\,1001\,0110 \cdots$$

is defined by

$$b = \sum b_i 2^i, \quad b_i \in \{0, 1\} \quad \Longrightarrow \quad t_b = \sum b_i \mod 2.$$

Definition 2.22. For a word u, let $\mathrm{first}(u)$ and $\mathrm{last}(u)$ denote the first and last letters of u, respectively. An *overlap* is a word of the form $uu\,\mathrm{first}(u)$ (or equivalently, $\mathrm{last}(u)uu$). A word w is *overlap-free* if it does not contain any overlaps.

Theorem 2.23 (Shelton and Soni [71]). *Let $a \geq 0$ and let μ be the Thue–Morse morphism (Definition 2.20). The words $\mu^a(00)$ and $\mu^a(001001)$ are overlap-free squares of lengths 2^{a+1} and $3 \cdot 2^{a+1}$, respectively.*[1]

Example 2.24 (Examples of Theorem 2.23.). The following overlap-free squares exemplify the first few possible lengths, 2, 4, 6, 8, and 12:

1 There is a minor typo in Shelton and Soni's paper (line 10 of page 98), equivalent to writing $\mu^a(001)$ instead of $\mu^a(001001)$.

$$\mathtt{00}, \quad \mu(\mathtt{00}) = \mathtt{0101}, \quad \mathtt{001001}, \quad \mu^2(\mathtt{00}) = \mathtt{01100110},$$
$$\mu(001001) = 010110010110.$$

Theorem 2.23 is used in the proof of Theorem 2.25.

Theorem 2.25 (Shelton and Soni [71]). *Let ℓ be a positive integer. The following are equivalent:*
1. *There exist an overlap-free binary word y and a word x such that y contains xx and $\ell = |xx|$;*
2. $\ell \in \{2^a : a \geq 1\} \cup \{3 \cdot 2^a : a \geq 1\}$.

Lemma 2.26. *If a cube www contains another cube xxx, then either $|x| = |w|$, or $xx\,\mathrm{first}(x)$ is contained in the first two consecutive occurrences of w, or $\mathrm{last}(x)xx$ is contained in the last two occurrences of w.*

Proof. We prove the contrapositive. Suppose $xx\,\mathrm{first}(x)$ is not contained in the first two consecutive occurrences of w and $\mathrm{last}(x)xx$ is not contained in the last two occurrences of w. Then the middle $\mathrm{last}(x)x\,\mathrm{first}(x)$ of the factor xxx has $\mathrm{last}(w)w\,\mathrm{first}(w)$ as a factor, and hence $|x| \geq |w|$. □

Theorem 2.27. *The deficiency of cube-free binary words is unbounded.*

Proof. Given k, we shall find a cube-free word x with $D(x) \geq k$. Pick a number n such that $2^n \geq 2k + 1$. Let $w := \mu^n(0)$, which is a word of length $\ell := 2^n$. By Theorem 2.23, ww is overlap-free. Let $x = ww\hat{w}$, where $\hat{w}$ is the proper prefix of w of length $|w| - 1$. By Lemma 2.26, x is cube-free. The complexity of x is at most $|w|$ as we can just make one loop of length w, with labels from $w_1, \ldots, w_\ell$. Hence,

$$\begin{aligned} D(x) &\geq \left\lfloor \frac{|x|}{2} \right\rfloor + 1 - |w| \geq \frac{|x|}{2} - |w| \\ &= \frac{3|w| - 1}{2} - |w| = \frac{|w|}{2} - \frac{1}{2} \geq k. \end{aligned}$$

□

Theorem 2.28 (Thue [79]). *The infinite Thue–Morse word (Definition 2.21) is overlap-free (Definition 2.22).*

Lemma 2.29. *Fix j and k and let t_x denote the xth bit of the Thue–Morse word. The function*

$$f(u) = t_{x(u)-1}, \quad \text{where } x(u) = 3^{k-j}(3u + 2),$$

is eventually nonconstant.

Proof. Gelfond [24] showed that $\mathbf{t}$ has no infinite arithmetic progressions (see also Morgenbesser, Shallit, and Stoll [57]). □

Lemma 2.30. *For each $k \geq 1$ there is a sequence $x_{1,k}, \ldots, x_{k,k}$ of positive integers such that*

$$\sum_{i=1}^{k} a_i x_{i,k} = 2\sum_{i=1}^{k} x_{i,k}, \quad a_i \in \mathbb{N} \quad \Longrightarrow \quad a_1 = \cdots = a_k = 2.$$

Let t_j denote bit j of the infinite Thue–Morse word. Then we can ensure that:
1. $x_{i,k} + 1 < x_{i+1,k}$ *and*
2. $t_{x_{i,k}} \neq t_{x_{i+1,k}}$ *for each* $1 \leq i < k$.

Proof. Let

$$x_{1,1} = 1.$$

Given $x_{1,k-1}, \ldots, x_{k-1,k-1}$, we let

$$x_{i,k} = 3x_{i,k-1}, \quad \text{for } i < k,$$
$$x_{k,k} = 3u_{k-1} + 2,$$

for a sufficiently large number u_{k-1}. Then we have

$$\sum_{i=1}^{k} a_i x_{i,k} = 2\sum_{i=1}^{k} x_{i,k} \Longrightarrow a_k \equiv 2 \pmod 3.$$

Thus, either $a_k = 2$ or $a_k \geq 5$. If $a_k \geq 5$, then

$$\sum_i a_i x_{i,k} \geq 5x_{k,k} = 15u_{k-1} + 10$$
$$> 6\sum_{i<k} x_{i,k-1} + 6u_{k-1} + 4 = 2\sum_{i<k} x_{i,k} + 2(3u_{k-1} + 2) = 2\sum_{i\leq k} x_{i,k},$$

provided

$$3u_{k-1} + 2 > 2\sum_{i<k} x_{i,k-1},$$

so we conclude $a_k = 2$. Then we can cancel a_k, divide by three, and reduce to the induction hypothesis.

Thus, our numbers are

$$x_{1,2} = 3, \quad x_{2,2} = 3u_1 + 2,$$
$$x_{1,3} = 3^2, \quad x_{2,3} = 3(3u_1 + 2), \quad x_{3,3} = 3u_2 + 2,$$

and in general

$$x_{j,k} = 3^{k-j}(3u_{j-1} + 2).$$

To ensure Item 1 we just take u_{j-1} sufficiently big. To ensure Item 2, we apply Lemma 2.29. □

Definition 2.31. We define a family of words DOF(d) parametrized by $d \in \mathbb{N}$. Let $k = 2d - 1$. For each $1 \le i \le k$ let $x_i = x_{k+1-i,k}$, where $x_{j,k}$ are as in Lemma 2.30. Note that since $x_{i,k} + 1 < x_{i+1,k}$, we have $x_i > x_{i+1} + 1$. Let DOF(d) $= w$, where

$$\begin{aligned} w &= \left(2\prod_{i=1}^{x_1-1} t_i\right)^2 t_{x_1}\left(2\prod_{i=1}^{x_2-1} t_i\right)^2 t_{x_2}\left(2\prod_{i=1}^{x_3-1} t_i\right)^2 \cdots t_{x_{k-1}}\left(2\prod_{i=1}^{x_k-1} t_i\right)^2 \\ &= \lambda_1 t_{x_1}\lambda_2 \cdots t_{x_{k-1}}\lambda_k, \end{aligned}$$

where $\lambda_i = (2\tau_i)^2$, $\tau_i = \prod_{j=1}^{x_i-1} t_j$, and t_i is the ith bit of the infinite Thue–Morse word on $\{0, 1\}$, which is overlap-free (Theorem 2.28).

Theorem 2.32. *For each $d \ge 1$, the word* DOF(d) *is overlap-free.*

Proof. Suppose a word uu is contained in w = DOF(d).

Proof that the number of 2s in uu is either zero or two. Let $o_1, \dots, o_{2a}$ denote the occurrences of 2s in uu and suppose $a \ge 1$. Let $\delta_i = o_{i+1} - o_i$. Then the sequence $(\delta_1, \dots, \delta_a)$ is an interval in the sequence

$$(x_1 - 1, x_1, x_2 - 1, x_2, \dots, x_{k-1} - 1, x_{k-1}, x_k - 1).$$

Since $x_i > x_{i+1} + 1$, in particular $|x_i - x_{i+1}| > 1$, this sequence is injective, i. e., no two entries are the same. But $(o_1, \dots, o_a) = (o_{a+1} - |u|, \dots, o_{2a} - |u|)$. So $\delta_{a+1} = o_{a+2} - o_{a+1} = o_2 - o_1 = \delta_1$, which implies $a = 1$.

So either Case 1 or Case 2 below is obtained.

Case 1: The number of 2s in uu is zero. Then certainly uu first(u) is not contained in w, since the infinite Thue–Morse word is overlap-free.

Case 2: The number of 2s in uu is two. Then we have one of the following two cases.

1. uu is contained in a word of the form

$$t_1 \cdots t_{x_i} \quad 2 \quad t_1 \cdots t_{x_{i+1}-1} \quad 2 \quad t_1 \cdots t_{x_{i+1}}.$$

   We guard against that by making sure that:
   - $t_{x_i} \ne t_{x_{i+1}-1}$ (Lemma 2.30) and
   - $2 \ne t_{x_{i+1}}$ (the Thue–Morse word uses only the letters 0 and 1).
2. uu is contained in a word of the form

$$t_1 \cdots t_{x_i-1} \quad 2 \quad t_1 \cdots t_{x_i} \quad 2 \quad t_1 \cdots t_{x_{i+1}-1}.$$

Since uu contains exactly two 2s and the t_j are not 2s, it follows that $uu = a2b2c$, where a, b, c are words over the binary alphabet $\{0, 1\}$. Then $u = a2b_1 = b_2 2c$, where $b = b_1 b_2$, so $a = b_2$ and $c = b_1$, so actually $u = a2c$ and $t_1 \cdots t_{x_i} = b = ca$. Here then $|ca| = x_i$. If $|a| \leq 2$, then consequently

$$x_i - 2 \leq |c| \leq x_{i+1} - 1,$$

which contradicts $x_{i+1} < x_i - 1$.

Now assume $|a| \geq 3$. The Thue–Morse word is a concatenation of disjoint occurrences of the words 01 and 10. Each of these two words is of the form $z\overline{z}$, where $\overline{z} = 1 - z$.

Let $u = t_1 \cdots t_{x_i - 1}2$ and $v = t_1 \cdots t_{x_i - 1} t_{x_i} 2$. If u has odd (even) length then u ends in, say, $z\overline{z}2$ ($z\overline{z}d2$), respectively, where d is a symbol. Then v ends in $z\overline{z}e2$ ($z\overline{z}d\overline{d}2$), respectively, where e is another symbol. Either way, we have inconsistent values for the last two symbols of a. □

Theorem 2.33. *The complexity deficiency of overlap-free words over an alphabet of size three is unbounded.*

Proof. Let $d \geq 1$. We will show that the word $w = \mathrm{DOF}(d)$ has deficiency $D(w) \geq d$. Let M be the NFA with code [69]

$$[+0^{x_1 - 1}]0[+0^{x_2 - 1}]0 \cdots 0 * [+0^{x_k - 1}],$$

where $*$ indicates the final state. Let $X = \sum_{i=1}^{k} x_i$. Then M has $k - 1 + X$ edges but only $q = X$ states, and w has length

$$n = k - 1 + 2X = 2(d - 1) + 2X,$$

giving $n/2 + 1 = d + X$.

Suppose v is a word accepted by M. Then M on input v goes through each loop of length x_i some number of times $a_i \geq 0$, where

$$k - 1 + \sum_{i=1}^{k} a_i x_i = |v|.$$

If additionally $|v| = |w|$, then by Lemma 2.30 we have $a_1 = a_2 = \cdots = a_k$, and hence $v = w$. Thus,

$$D(w) \geq \lfloor n/2 + 1 \rfloor - q = d + X - X = d.$$

By Theorem 2.32, w is overlap-free. □

Lemma 2.34. *Let x with $|x| = n$ be uniquely accepted by an NFA M, along a unique walk of states $w = (q_0, q_1, \dots, q_n)$. For any t, t_1, t_2, and r_i, s_i with*

$$(p_1, r_1, \dots, r_{t-2}, p_2) = (q_{t_1}, \dots, q_{t_1+t})$$

and

$$(p_1, s_1, \dots, s_{t-2}, p_2) = (q_{t_2}, \dots, q_{t_2+t}),$$

we have $r_i = s_i$ for each i.

Proof. Suppose that for some i, $r_i \neq s_i$. Then we can replace the segment $(r_1, \dots, r_{t-2})$ by $(s_1, \dots, s_{t-2})$ in w to obtain a second accepting walk w'. □

Note that in Lemma 2.34, it may very well be that $t_1 \neq t_2$.

Theorem 2.35. *Overlap-free binary words have deficiency bound* 1.

Proof. Suppose w is a word satisfying $D(w) \geq 2$ and consider the sequence of states visited in a witnessing computation. As in the proof of Theorem 2.45, either there is a state that is visited four times, and hence there is a cube in w, or there are three *state cubes* (states that are visited three times each), and hence there are three squares in w. By Theorem 2.25, an overlap-free binary word can only contain squares of length 2^a or $3 \cdot 2^a$, and hence can only contain powers u^i, where $|u|$ is of the form 2^a or $3 \cdot 2^a$ and $i \leq 2$.

In particular, the length of one of the squares in the three state cubes must divide the length of another. So if these two state cubes are disjoint, then the shorter one repeated can replace one occurrence of the longer one, contradicting Lemma 2.34.

So suppose we have two state cubes, at states p_1 and p_2, that overlap. At p_1 then we read consecutive words ab that are powers $a = u^i$, $b = u^j$ of a word u, and since there are no cubes in w we must have $i = j = 1$, so actually $a = b$. At p_2 we have words c, d that are powers of a word v and again the exponents are 1 and $c = d$.

The overlap means that in one of the two excursions of the same length starting and ending at p_1, we visit p_2. By the uniqueness of the accepting walk we then visit p_2 in both of these excursions. If we suppose the state cubes are chosen to be of minimal length, then we only visit p_2 once in each excursion. If we write $a = rs$, where r is the word read when going from p_1 to p_2 and s is the word going from p_2 to p_1, then $c = sr$ and w contains $rsrsr$. In particular, w contains an overlap. □

Remark 2.36. In computability theory, the effective Hausdorff dimension dim and effective packing dimension Dim of a single infinite binary sequence $\mathbf{u}$ are defined and related to Kolmogorov complexity C. It is shown (see [18, Theorem 13.3.4 and Corollary 13.11.12]) that

$$\dim(\mathbf{u}) = \liminf_n \frac{C(u_1 \cdots u_n)}{n} \quad \text{and} \quad \mathrm{Dim}(\mathbf{u}) = \limsup_n \frac{C(u_1 \cdots u_n)}{n}.$$

These results, together with the idea that automatic complexity is a miniaturization of Kolmogorov complexity, constitute our motivation for Definition 2.37 and Definition 2.39.

Definition 2.37. For an infinite word $\mathbf{u}$ define the *deterministic automatic Hausdorff dimension* of $\mathbf{u}$ by

$$I(\mathbf{u}) = \liminf_{u \text{ prefix of } \mathbf{u}} \frac{A(u)}{|u|}.$$

Define the *deterministic automatic packing dimension* of $\mathbf{u}$ by

$$S(\mathbf{u}) = \limsup_{u \text{ prefix of } \mathbf{u}} \frac{A(u)}{|u|}.$$

The connection between effective dimension and automatic dimension is not merely by analogy, as Theorem 2.38 shows.

Theorem 2.38. *If* $\mathbf{x}$ *is an infinite word with* $\dim(\mathbf{x}) > 0$, *then* $I(\mathbf{x}) > 0$.

Proof. This follows from the Kolmogorov complexity calculation in [69, Theorem 9]. □

For nondeterministic complexity, in light of Theorem 2.2 it is natural to establish the following definition.

Definition 2.39. Define the *nondeterministic automatic Hausdorff dimension* of $\mathbf{u}$ by

$$I_N(\mathbf{u}) = 2 \cdot \liminf_{u \text{ prefix of } \mathbf{u}} \frac{A_N(u)}{|u|}$$

and define S_N analogously.

Theorem 2.40 (Shallit and Wang's Theorem 18). *We have* $\frac{1}{3} \le I(\mathbf{t}) \le S(\mathbf{t}) \le \frac{2}{3}$.

We are now ready to strengthen Theorem 2.40.

Theorem 2.41. *We have* $I(\mathbf{t}) = \frac{1}{2}$ *and* $I_N(\mathbf{t}) = S_N(\mathbf{t}) = 1$.

Proof. The inequality $I(\mathbf{t}) \ge \frac{1}{2}$ and the fact that $I_N(\mathbf{t}) = S_N(\mathbf{t}) = 1$ follow from the observation that the proof of Theorem 2.35 applies equally for deterministic complexity. The inequality $I(\mathbf{t}) \le \frac{1}{2}$ was already implicit in the proof of [69, Theorem 18]. Let $T(m) = t_0 \cdots t_{m-1}$. In the table they give, with $m = 2^{2n+1}$, we read off the inequality $A(T(m)) \le m + 3 - 2^{2n} = \frac{m}{2} + 3$. □

2.2.4 Almost square-free words

Definition 2.42 (Fraenkel and Simpson [20]). A word over the alphabet $\{0, 1\}$ whose square factors all belong to the set $\{00, 11, 0101\}$ is called *almost square-free*.

A generalization to arbitrary alphabets: A word over the alphabet Σ whose square factors are either of the form aa, $a \in \Sigma$, or $abab$, $a, b \in \Sigma$, and such that for all $a, b \in \Sigma$, at most one of $abab$ and $baba$ occurs in w is called *almost square-free*. It appears that this generalization applies to Theorem 2.45.

Definition 2.43 ([76]). The Van Eck sequence $a : \mathbb{N} \to \mathbb{N}$ is defined by $a_0 = 0$, and each a_{n+1} is either 0, if $a_n \neq a_j$ for all $j < n$, or $n-j$ for the largest $j < n$ with $a_n = a_j$, otherwise, and $a(n) = a_n$.

As the reader can verify, the longest prefix of the Van Eck sequence consisting of only single decimal digit numbers is

$$0010202216050265405303290 4936.$$

Theorem 2.44. *The Van Eck sequence is overlap-free.*

Proof. (We adapt Van Eck's proof that his sequence has infinitely many zeros.) Suppose there is an overlap: $a_r \dots a_{r+p-1}$ is equal to $a_{r+p} \dots a_{r+2p-1}$ and $a_r = a_{r+2p}$ as well. Assume this is the first overlap in the Van Eck sequence. Let $z = a_{r+p-1}$. Then by assumption $z = a_{r+2p-1}$, so the minimal $q > 0$ such that $z = a_{r+p-1+q}$ exists and satisfies $q \leq p$. Then by the definition of the Van Eck sequence, $a_{r+p+q} = q$. By assumption of overlap, $a_{r+q} = q$ as well, and $a_{r+q-1} = a_{r+q-1+p} = a_{r+p-1}$. By the definition of Van Eck applied to $a_{r+q} = q$, $a_{r+q-1} = a_{r-1}$. Therefore, $a_{r+p-1} = a_{r-1}$, so our overlap was not the first in the Van Eck sequence after all, that is, unless the Van Eck sequence starts with an overlap for block length p. But the Van Eck sequence starts with $(0, 0, 1)$, so $p \neq 1$, and the $(0, 0)$ never reappears, so $p \not\geq 2$. □

Theorem 2.45. *A word that is almost square-free has a deficiency bound of* 1.

Proof. First, let w be a word of length $n \leq 3$. Since $A_N(w) \geq 1$, w has deficiency $\lfloor n/2 \rfloor + 1 - A_N(w) \leq \lfloor n/2 \rfloor \leq 1$.

It remains to consider the case $|w| \geq 4$. Suppose w is a word of length $n \in \{2k, 2k+1\}$, where $k \geq 2$, with deficiency at least 2. Then there are $q = k - 1 \geq 1$ states occupied at $n + 1$ times. So $n + 1 \in \{2k + 1, 2k + 2\} = \{2q + 3, 2q + 4\}$ times. There are at least $2q + 3$ times and only q states, so by the extended pigeonhole principle (Theorem 2.7), we are in one of the following three cases:

- Case 1. There is at least one state that is visited at least 5 times. Then by Theorem 2.5, *w contains a 4-power.*
- Case 2. There is at least one state p_1 that is visited at least 4 times and another state $p_2 \neq p_1$ that is visited at least 3 times. Then by Theorem 2.5, there are a cube xxx and a square yy in w. Since *w has no squares of length* > 4, we must have $|xx| \leq 4$ and $|yy| \leq 4$ and hence $1 \leq |x| \leq 2$ and $1 \leq |y| \leq 2$. We next consider possible lengths of x and y.

- Subcase $|x| = 2$. Consider $x = ab$, where $|a| = |b| = 1$. If $a \neq b$, then $xxx \in \{101010, 010101\}$, so 1010 occurs in w and w is not almost square-free. If $a = b$, then 0000 or 1111 occurs in w, and again w is not almost square-free.
- Subcase $|x| = 1$, $|y| = 2$. In this case, the xxx and yy occurrences must be disjoint, because the states in a yy occurrence are $p_2p_3p_2p_3p_2$ for some p_3 which must be disjoint from $p_1p_1p_1p_1$ when $p_1 \neq p_2$. But then we can replace these by $p_2p_3p_2p_3p_2p_3p_2$ and p_1p_1, respectively, giving two distinct state sequences leading to acceptance, contradicting Lemma 2.34.
- Subcase $|x| = 1$, $|y| = 1$. In this case again the occurrences of xxx and yy must be disjoint, since $p_1 \neq p_2$. We can replace p_1^4 and p_2^3 by p_1 and p_2^6, respectively, again contradicting Lemma 2.34.

- Case 3. There are at least three states p_1, p_2, p_3 (all distinct) that are each visited at least 3 times. Then by Theorem 2.5, there are three squares u_iu_i at three distinct states p_i, $1 \leq i \leq 3$. By assumption, $|u_iu_i| \leq 4$, so $|u_i| \leq 2$.
  - Subcase 3.1. $|u_i| = |u_j| = 1$ for two values $1 \leq i < j \leq 3$. The argument is entirely analogous to that in Case 2.
  - Subcase 3.2 $|u_j| = |u_k| = 2$ for two values $1 \leq j < k \leq 3$.
    - Subsubcase 3.2.1. If disjoint, we can replace u_j^2 by u_k^2 to get u_k^4, again a *4-power*, by the argument of Subcase 3.1.
    - Subsubcase 3.2.2. If nondisjoint with full overlap, then

$$p_ja_1p_ja_2p_j$$

and

$$p_kb_1p_kb_2p_k$$

become

$$p_jp_kp_jp_kp_jp_k,$$

and immediately we get 10101 *or* 01010 *or a 4-power in w.*
    - Subsubcase 3.2.3. In the case of partial overlap only, $p_ja_1p_ja_2p_j$ and $p_kb_1p_kb_2p_k$ become, by Lemma 2.34, $p_jap_jap_j$ and $p_kbp_kbp_k$ and then

$$p_jap_jp_kp_jp_kbp_k.$$

By Lemma 2.34 again, this must be

$$p_jp_kp_jp_kp_jp_kp_jp_k = (p_jp_k)^4,$$

so the read word must be of the form *abababa*, giving *an occurrence of* 1010 *(if $a \neq b$) or of a 7-power (if $a = b$) in w.*

Thus, all cases are covered, and the result is proved. □

Corollary 2.46. *There is an infinite binary word having a hereditary deficiency bound of* 1.

Proof. We have two distinct proofs. On the one hand, Fraenkel and Simpson [20] show there is an infinite almost square-free binary word, and the result follows from Theorem 2.45. On the other hand, the infinite Thue–Morse word is overlap-free (Theorem 2.28) and the result follows from Theorem 2.35. □

2.3 Open problems

Question 2.1. Does there exist a short proof of the statement that no partial DFA with four states uniquely accepts the word 0001101? (Part of the task here is to define "short proof.")

Question 2.2. Is the Van Eck sequence almost square-free?

Question 2.3. Is there an infinite binary word having hereditary deficiency 0?

Remark 2.47. We obtained some numerical evidence that the answer to Question 2.3 is "yes." For instance, we found that there are 108 binary words of length 18 having hereditary deficiency 0.

2.4 Exercises

Exercise 2.1. Prove that the equation $143x + 91y + 77z = 2692$ has the unique solution $(x, y, z) = (6, 10, 12)$ over $\mathbb{N}$.

A complete semiformal proof of Exercise 2.1 is available at the following URL.

https://github.com/bjoernkjoshanssen/ac-exercises/

Exercise 2.2. Show that there is no square-free binary word of length 4.

A complete proof of Exercise 2.2 is available at the following URL.

https://github.com/bjoernkjoshanssen/ac-exercises/

Exercise 2.3. Define walks in labeled digraphs inductively. Define unlabeled walks inductively from labeled walks. Use this to define A_N without directly referring to NFAs. Use this to show that the two-state Kayleigh digraph uniquely accepts a given word of length 3, $x = x_0x_1x_2$.

A complete solution to Exercise 2.3 is available at the following URL.

https://github.com/bjoernkjoshanssen/ac-exercises/

Exercise 2.4. Define E_N formally and prove that $E_N(\varepsilon) \leq 0$, $E_N(\mathtt{0}) \leq 1$, and $E_N(\mathtt{00}) \leq 1$.

A complete solution to Exercise 2.4 is available at the following URL.

https://github.com/bjoernkjoshanssen/ac-exercises/

Exercise 2.5. Show that a binary word that is (abstractly) almost square-free is generalized almost square-free.

Exercise 2.6. Show that the word `000` is almost square-free but not overlap-free.

Exercise 2.7. Show that the word `011011` is not almost square-free. (Note, but do not show formally, that the word `011011` is, however, overlap-free.)

Complete solutions to Exercise 2.5, Exercise 2.6, and Exercise 2.7 are available at the following URL.

https://github.com/bjoernkjoshanssen/ac-exercises/

3 Edge complexity and digraphs

3.1 Edge-counting automatic complexity

Shallit and Wang counted states in their definition of automatic complexity, although for (complete) DFAs one could equivalently count edges (Theorem 3.2). For NFAs, edge counting is worth considering separately from state counting and is preferable in some ways. For instance, the edge complexity of a word w is bounded below by the number of distinct symbols of w, and the case where this bound is an equality is interesting.

Definition 3.1. We define three notions of edge complexity for a word x, analogous to A_N, A^-, and A.

- The nondeterministic edge complexity $E_N(x)$ is the minimum number of edges in the digraph of an NFA that uniquely accepts x, i. e., accepts x along only one walk, and accepts no other word of length $|x|$.
- The deterministic incomplete edge complexity $E^-(x)$ is the minimum number of edges in the digraph of a deterministic NFA that uniquely accepts x.
- The deterministic complete edge complexity $E(x, \Sigma)$ is the minimum number of edges in the digraph of a DFA over the alphabet Σ that uniquely accepts x.

Clearly, $E_N(x) \leq E^-(x) \leq E(x, \Sigma)$. In fact, E is not really anything new.

Theorem 3.2. *We have $E(x, \Sigma) = |\Sigma| \cdot A(x, \Sigma)$ for each x and Σ with $x \in \Sigma^*$.*

Proof. By definition, a DFA over Σ with its set of states Q has $|\Sigma| \cdot |Q|$ edges. □

If a morphism sends all symbols of Σ to words of length k, then the edge-automatic complexity satisfies $E_N(\pi x) \leq kE_N(x)$. For comparison, consider Lemma 3.3, which was used in [37].

Lemma 3.3. *Let A denote deterministic automatic complexity. Let $\pi : \{0,1\}^* \to \{0,1\}^*$ be an injective homomorphism with $|\pi(0)| = |\pi(1)|$. Then $A(x) \leq A(\pi(x))$ for each word x.*

Proof. Let M be a witnessing automaton for $A(\pi(x))$, with transition function δ. We can now make an automaton M' with the same states as M (and the same initial and final states) that uniquely accepts x among words of length $|x|$ as follows. Throw out all the edges of M. Put an edge labeled i from q_1 to q_2 in M' if $\delta(q_1, \pi(i)) = q_2$ in M.

It is clear that M' accepts x. We now turn to uniqueness.

Suppose $|y| = |x|$ and M' accepts y. Since $|\pi(0)| = |\pi(1)|$, $|\pi(y)| = |\pi(x)|$. But M only accepts one word of length $|\pi(x)|$, so we must have $\pi(y) = \pi(x)$. Since π is injective, it follows that $y = x$. □

The number of distinct symbols $\mathrm{wd}(x)$ occurring in a word x is the cardinality of the range of x, viewed as a function with domain $|x|$, and satisfies $\mathrm{wd}(x) \leq E_N(x)$.

https://doi.org/10.1515/9783110774870-003

Theorem 3.4. *For each Σ and $x \in \Sigma^*$, $E_N(x) \le E^-(x) \le |x|$. The bound is sharp as Σ varies: for each $n \in \mathbb{N}$ there exist Σ and $x \in \Sigma^n$ with $E_N(x) = |x|$.*

Proof. Let x be a word with $\mathrm{wd}(x) = |x|$. For $|x| = 0, 1, 2, 3$, we may write $x = \varepsilon, \mathtt{0}, \mathtt{01}, \mathtt{012}$. Then

$$|x| = \mathrm{wd}(x) \le E_N(x) \le |x|.$$

□

We have $E^-(\varepsilon) = 0$, $E^-(\mathtt{0}^n) = 1$ for all $n \ge 1$, $E^-(01) = 2$ by the above, $E^-(010) = 2$, and $E^-(011) = 2$. Moreover, $E^-(0^{n-1}1) = 2$ for all $n \ge 2$. Indeed, the upper bound is easy, and the lower bound comes from the number of distinct symbols occurring in 001.

Theorem 3.5. *Let $G = (V, E)$ be a connected undirected graph. Then $|V| \le |E| + 1$.*

Proof. It suffices to show that there exists an ordering of the vertices of V, $v_0, \dots, v_{q-1}$, where $q = |V|$, such that for each $1 \le i \le q-1$ there is an edge e_i connecting v_i to some v_j, $j < i$. Then these edges are distinct: whenever $j < i$, e_j connects v_j to some v_k, $k < j$, and hence v_i is not an endpoint of e_j, whereas e_i connects v_i to some v_ℓ, $\ell < i$.

Let v_0 be an arbitrary vertex of V. Let w_1 be an arbitrary vertex of V such that $w_1 \ne v_0$. Since G is connected, there is a walk W in G connecting v_0 to w_1. Since the first vertex in W is v_0 and the last vertex in W is not v_0, there must be some edge in W that connects v_0 to some edge $v_1 \ne v_0$; call it e_1.

Inductively, if $S = \{v_0, \dots, v_{i-1}\}$ are given, let w_i be any vertex not equal to any v_j, $j < i$. By connectedness there is a walk from, say, v_0 to w_i. Since the first vertex in this walk is in S and the last one is not, there must be some edge e_i in W connecting some vertex in S to some vertex v_i not in S. See Exercise 4.2. □

Theorem 3.6. *For all words x, $E^-(x) \ge A^-(x) - 1$ and $E_N(x) \ge A_N(x) - 1$.*

Proof. For any connected NFA M, let $q(M)$ be its number of states and $e(M)$ its number of edges. By Theorem 3.5,

$$\min\{e(M) : M \text{ uniquely accepts } x\} \ge \min\{q(M) : M \text{ uniquely accepts } x\} - 1.$$

□

Theorem 3.7. *For all words x, $E^-(x) \le 2A^-(x) - 1$ and $E_N(x) \le 2A_N(x) - 1$.*

Proof. At most four edges can be incident to any given vertex, as shown in [69]. Moreover, there cannot be four edges incident to the starting vertex. □

There need not be another vertex, besides the starting vertex, that does not have four edge tips incident to it, as Kayleigh graphs show.

We now have two proofs that Hyde's Theorem 2.2 is sharp: via edge complexity (Theorem 3.4) and via square-free words (Theorem 2.8). We cannot combine them, in the sense that a square-free word x need not have $E_N(x) = |x|$. A counterexample is $x = \mathtt{010}$, which has $E_N(x) = A_N(x) = 2$. On the other hand, a square-free word has $A_N(x) = \lfloor \frac{|x|}{2} \rfloor + 1$ and hence $E_N(x) \ge \lfloor \frac{|x|}{2} \rfloor$. In fact, we have the following theorem.

Theorem 3.8. *The following are equivalent for a word x:*
1. $E_N(x) = |x|$;
2. $E^-(x) = |x|$;
3. $E^\pm(x) = |x|$;
4. *x has maximal range.*

Proof. The nontrivial implication is: $E^\pm(x) = |x| \implies x$ has maximal range. So suppose the range of x is nonmaximal; thus, $x = uavaw$ for some words $u, v, w \in \Sigma^*$ and for some $a \in \Sigma$. Then we form a deterministic NFA M which proceeds along a straight path for uav, then reuses the a edge, and then continues until, if ever, $v(k) \neq w(k)$ for some k, at which point M branches off. This ensures determinism. To ensure bi-determinism, add the assumption that the first occurrence of a in $x = uavaw$ is the first occurrence in x of any symbol appearing more than once in x. □

Theorem 3.8 is optimal with respect to the master diagram (4.1) in that the implication ($A^{\mathrm{nb}}(x) = |x| \implies x$ has maximal range) does not hold. For a counterexample, consider any word of the form $0^{n-1}1$, $n \geq 1$.

By Theorem 3.8 the property of having maximal edge complexity is polynomial time computable. For a follow-up, see Question 3.2.

Theorem 3.9 is an extended version of the phenomenon we saw at length 6 with three states in Theorem 2.19.

Theorem 3.9. *Given $k \geq 1$, define $x = 0^k 1^{k+1} 0^{k-1}$. For each $k \geq 2$, $(A_N(x), E_N(x)) \leq (k+1, k+2)$.*

Proof. Consider the following NFA M:

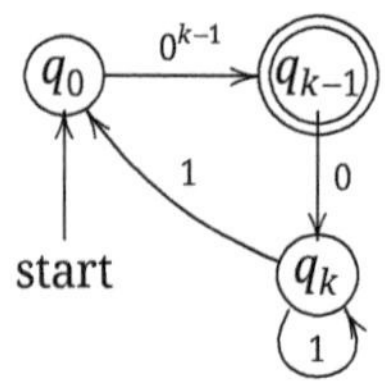

A word of length $3k$ accepted by M clearly has 0^k as a prefix and 10^{k-1} as a suffix (final segment). The word y in the middle of length k must start and end in q_k. Since the big cycle has length $k+1 > k$, we must have $y = 1^k$ and the unique walk is determined. The witnessing NFAs for $k = 3$ and $k = 4$ are shown in (3.1) and (3.2), respectively:

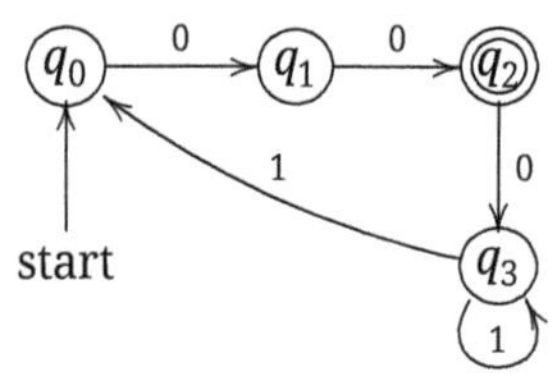

(3.1)

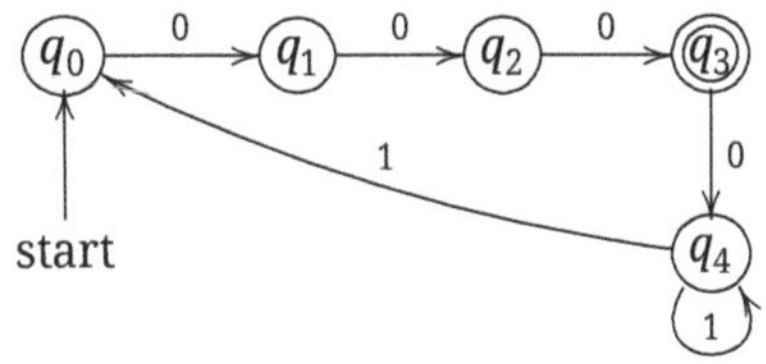

(3.2)

In general, given k let j be the number of times the accepting state q_2 is visited in an accepting run of x. We argue for uniqueness of this accepting run and the fact that we must have $j = 2$ as follows:

- Case $j = 1$: The NFA accepts a word of length $k - 1$, namely, 0^{k-1}, and $k - 1 < 3k$.
- Case $j = 2$: The NFA accepts the words $0^k 1^b 1 0^{k-1}$ for any $b \in \mathbb{N}$. These words have length $2k + b$. With $b = k$, the length is therefore $3k$.
- Case $j \geq 3$: The NFA accepts words of the form $0^k 1^{b_1} 1 0^k 1^{b_2} 1 0^{k-1} \cdots$ which have length at least $3k + 1$. □

Theorem 3.10. *Let $k \in [3, 12]$ and $x = 0^k 1^{k+1} 0^{k-1}$. Then*

$$(A_N(x), E_N(x), A^-(x), E^-(x)) = (k + 1, k + 2, k + 2, k + 3).$$

We have verified Theorem 3.10 by computation.

Theorem 3.11 is a bit surprising for the following reason. By definition, nondeterminism involves having at least two edges available and only using one (at a time). For Theorem 3.11, we need an example where dropping determinism allows us to use *fewer* edges.

Theorem 3.11. *We have $E_N \neq E^-$.*

Proof. Using computer search, we find that there is no binary example of length 7. However, the result follows from Theorem 3.10. □

Remark 3.12. Results like Theorem 3.11 are *weak* separations, in that they do not show that the complexity notions involved differ on infinitely many inputs. Thus, we have limited understanding, but reasonable computational prowess.

A strong separation of A^- and E_N can be given as follows: if x is a rainbow word, $E_N(x) = |x|$, but $A^-(x) \leq \frac{|x|}{2} + 1$.

Similarly, $A(0^n) = 2 > 1 = A_N(0^n)$ for $n \geq 1$, so we separate A and A^- in the stronger sense. On the other hand, A_N and A^- are asymptotically equal in the sense that for each q, as $|x| \to \infty$, $A_N(x) = q$ implies $A^-(x) = A_N(x)$. This potentially makes it hard to show that they are strongly different.

Remark 3.13. Theorem 3.10 cannot be generalized to all large k. Consider Shallit and Wang's Lemma 14. Let $(n_1, n_2, n_3) = (7, 11, 13)$ and $P = n_1 n_2 n_3 = 1001$. The equation

$$\frac{P}{n_1}x + \frac{P}{n_2}y + \frac{P}{n_3}z = 143x + 91y + 77z = 143 \times 6 + 91 \times 10 + 77 \times 12$$

has a unique solution $(x, y, z) = (n_1 - 1, n_2 - 1, n_3 - 1) = (6, 10, 12)$ by Exercise 2.1. The number of states would be $143 + 1 + 91 + 1 + 77 = 313 < n/3$, where n is the length of the accepted word. Of course,

$$0^{143\times6}1^{1+91\times10}0^{1+77\times12} = 0^{858}1^{911}0^{925}$$

is not quite of the form $0^k1^{k+1}0^{k-1}$, but it may be close enough. If we rearrange to put the shortest segment in the middle, $0^{911}1^{858}0^{925}$, then with $k = 857$ we have, with $z = 0^{54}0^k1^{k+1}0^{k-1}0^{69}$, by the subword inequality, $A^-(0^k1^{k+1}0^{k-1}) \leq A^-(z) \leq 313$, which is much less than $k + 1$.

We can certainly try to optimize and reduce the constant $k = 857$ here.

With $(n_1, n_2, n_3) = (4, 5, 7)$ we have $n = 35\times3+1+28\times4+1+20\times6 = 105+1+112+1+120$, so we can take $k = 104$ and the number of states is $35+1+28+1+20 = 85$, which fortunately is below $k+1$. The choice $(4, 5, 7)$ is actually optimal, i. e., it minimizes k for this construction based on Shallit and Wang's Lemma 14 and gives $k = 104$ and $q = 85$. Another search, not using that lemma, showed however that the equation $19x + 23y + 24z = 217$ can be used and gives $q = k = 68$.

Whereas most of our results on the automatic complexity of infinite families are upper or lower bounds, Theorem 3.14 is an example of an exact result.

Theorem 3.14. *We have $E^-(x) = E_N(x) = 3$, where $x = 0^{n-2}1b$ for all $n \geq 4$ and all symbols b.*

Proof. To show $\geq$, note that if only two edges are used, then since the `1` edge must be distinct from the `0` edge, the first two `0`s must use the same edge, which must hence be a loop. Then to maintain uniqueness the `1` edge must go to a new state. But then the last b cannot use either edge.

To show $\leq$, a simple construction suffices. □

Sometimes A_N and E_N do not agree on which of two words is the most complex.

Theorem 3.15. *There exist words x, y such that $E_N(y) < E_N(x)$ and $A_N(x) < A_N(y)$.*

Proof. See the following table:

	x	y
A_N	4	5
E_N	6	5

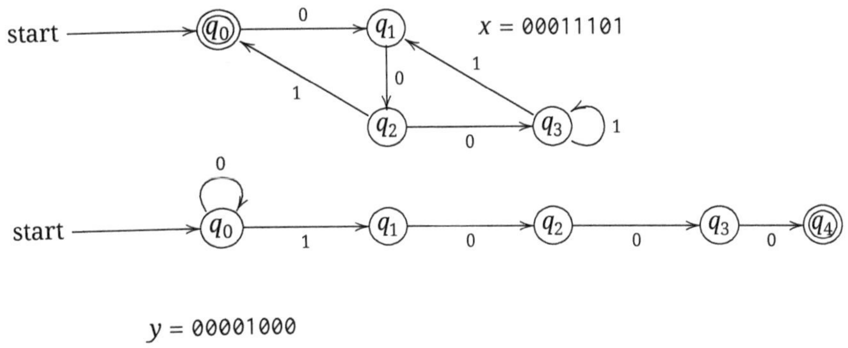

(3.3)

State counting, $0^3 1^3 01$ is simpler: one of those stacked-triangles examples, which is “almost” a Kayleigh graph layout, and Kayleigh graphs have a lot of edges. But edge counting, $0^4 10^3$ is simpler: it starts with slightly more 0s that are exploited. □

Theorem 3.16. *There exists a binary word x of length 7 such that there is no single NFA that witnesses both $E_N(x)$ and $A_N(x)$. There is no such word of length less than 7.*

Proof. Let $x = 0001101$ (which also arises in the study of counter automatic complexity, as an example with A^- exceeding the Hyde bound, and in the discussion after Theorem 3.11). The witnesses are 01234234 (five states, five edges), 01233120 (four states, six edges), and 00001234 (five states, five edges). □

3.2 Power-bordered words

Automatic complexity may be computationally hard in general. Still, looking at binary words of length 6 or less, it is not too hard to get a feel for what the A_N-complexity will be in each case. The only exception, in the author’s opinion, is that of power-bordered words.

Definition 3.17. A word is *power-bordered* if it is of the form xab^kcx, where ($0^0 = 0$ and) $|a|, |c| > 0$, $k|b| < |x| + |a| + |c|$, and

$$|xac| + ((|b| - 1) \vee 0) < \left\lfloor \frac{|xab^kcx|}{2} + 1 \right\rfloor. \tag{3.4}$$

Definition 3.17 is strange, but it does capture the words for which Theorem 3.18 is nontrivial. Inequality (3.4) guarantees the nontriviality. As an example, if $|x| = |a| = |b| = |c| = 1$ and $k = 2$, then xab^kcx is power-bordered.

The proof of Theorem 3.9 shows the more general Theorem 3.18.

Theorem 3.18 (Power-bordered theorem). *For any word y of the form xab^kcx, where $|a|, |c| > 0$ and $k|b| < |x| + |a| + |c|$, we have*

$$(A_N(y), E_N(y)) \le (|x| + |a| + |c| + ((|b| - 1) \vee 0))(1, 1) + (0, 1).$$

Here we order pairs (a, b) by the product order and $\vee$ denotes maximum.

A generalization of the power-bordered theorem (Theorem 3.18) inspired by Example 1.66 is given below:

Theorem 3.19. *Let u, v, y, z, w be words over an alphabet Σ. Let $x = uv(yz)^k ywu$, where $k \in \mathbb{N}$, and write $a = |zy|$ and $b = |wuvy|$. If*

$$\{(x_1, x_2) \in \mathbb{N}^2 \mid ax_1 + bx_2 = |x| - |uvywu| \text{ and } (x_1 > 0 \implies x_2 > 0)\}$$

is a singleton $(x_1, 1)$, then $A_N(x) \le |uvyzw|$.

Proof. Consider the NFA M in (3.5):

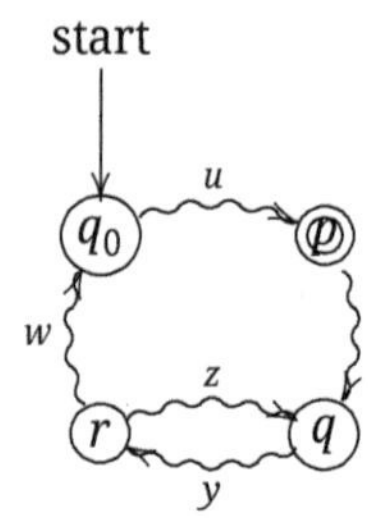

(3.5)

By convention, if $z = \varepsilon$, then $q = r$, forming a cycle. The language accepted by M is

$$u(vy(zy)^* wu)^* = \bigcup_{l \in \mathbb{N}} S_l,$$

where

$$S_l = u(vy(zy)^* wu)^l.$$

In order to show $\Sigma^{|x|} \cap \bigcup_l S_l = \{x\}$, note that $S_0 = \{u\}$ and $x \ne u$, since if $x = u$, then by definition of x this implies $x = \varepsilon$. Our equation becomes $0x_1 + 0x_2 = 0$ with $x_1 > 0 \implies x_2 > 0$. However, this has many solutions, not just one, so our assumptions are not satisfied.

If $l \ge 2$ and $x \in S_l$, then our equation has a solution with $x_2 \ge 2 > 1$, contrary to assumption.

Finally, with $l = 1$, we get $x_2 = 1$, and by assumption there is only one choice for x_1, giving $S_1 \cap \Sigma^{|x|} = \{x\}$. □

Theorem 3.20 (Deterministic power-bordered theorem). *For any word y of the form xab^kcx, where $|a|, |c| > 0$ and $k|b| < |x| + |a| + |c|$,*

$$(A^-(y), E^-(y)) \leq (|x| + |a| + |c| + ((|b| - 1) \vee 0) + \min\{|b|, |c|\})(1,1) + (0,1).$$

Proof. If $c = uv$, where $b = uw$ and $v(1) \neq w(1)$ (v and w immediately differ) or $v = w = \varepsilon$, then we have $y = xa(uw)^k uvx = xau(wu)^k vx$. This can be exploited at the cost of adding $|u| \leq \min\{|b|, |c|\}$ states and edges. □

3.3 One-cycle-free-path automata

The digraph G of an NFA generated from an accepting state sequence is a *unilaterally connected* digraph.

Definition 3.21. A digraph is unilaterally connected if for any pair of vertices u and v, either u is reachable from v or v is reachable from u. A directed spanning walk in a digraph G is a walk visiting every vertex of G.

For finite digraphs, unilateral connectivity is equivalent to the existence of a directed spanning walk [13, Exercise 7.1.3].

It is a concept intermediate between strongly connected (directed walks exist between any two vertices in any direction) and weakly connected (also known as "connected"). In automatic complexity theory we are interested in the case where there is a unique directed spanning walk of a given length between given vertices. However, we want a little more: not just a directed vertex-covering walk, but a directed edge-covering walk, i. e., a walk visiting every edge [21]. It is not the same thing, as the digraph (3.6) exemplifies:

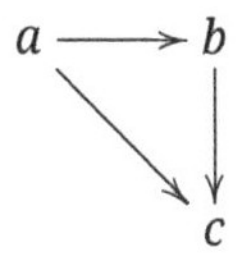

(3.6)

Union-free languages are regular languages defined by regular expressions not containing the union symbol ∪. They were studied by Nagy in 2006 [59]. Associated automata are one-cycle-free-path (1cfpa): there is only one cycle-free path from any given state to the final state. See also [10]. Note that the digraph of a 1cfpa with a single start state is, or may be assumed to be after removing states that do not affect the language, unilaterally connected. The converse fails, which indicates that our attention should focus on 1cfpa rather than (edge-)unilateral connectivity. In (3.7) we have a digraph that is edge-unilaterally connected but not 1cfpa: there is a walk from a to b covering every edge, but there is more than one cycle-free path from d to b.

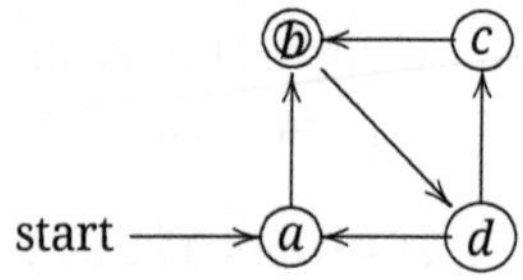

(3.7)

Nagy's example of a 1cfpa, with the "backbone" indicated by double arrows, is shown in (3.8):

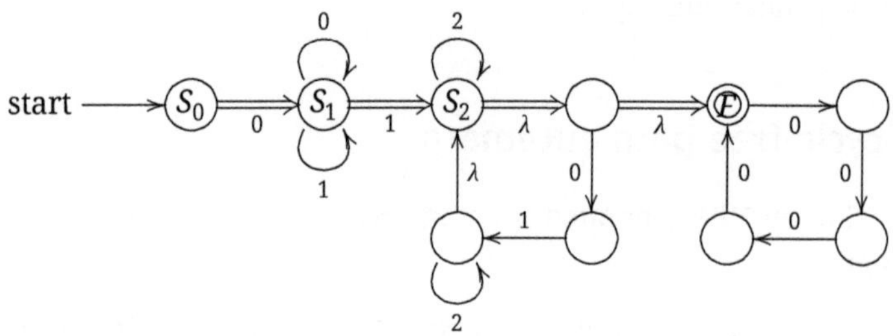

(3.8)

One difference with A_N witnesses is that the latter have degree at most 4 at each vertex. In Nagy's example, S_1 has degree 6 and S_2 has degree 5. In addition to the degree bound, A_N witnesses do not contain $(u^*v^*)^* = (u+v)^*$, where u and v are regexes, because that would violate uniqueness.

3.4 Truncated complexity

A_N witnesses have the extra property that there is a unique shortest walk from the start state to the end state that visits every edge. This walk is obtained from the witnessing walk by truncating any powers. For example, the word

$$10011010101000000001011$$

has the following witnessing walk for a set of states $Q = \{0, 1, 2, 3, 4, 5, 6, 7, 8, 9, A, B\}$:

$$012345623478989898989AB01,$$

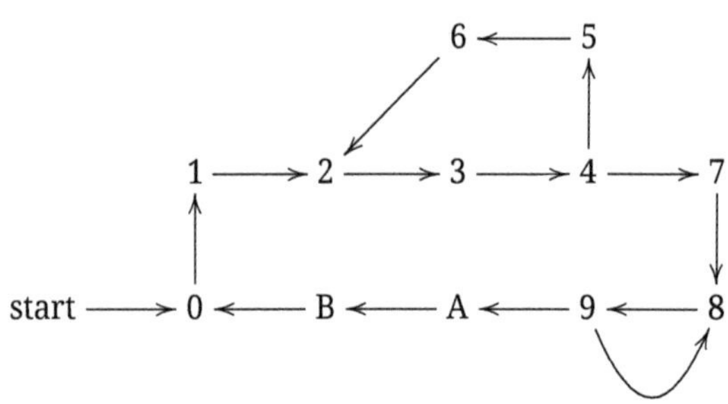

The truncated form of this walk is

$$012345623478989AB01.$$

The length of this truncated walk equals or exceeds the number of edges and could be its own complexity measure, wedged between E_N and A^{nb}.

Definition 3.22. The truncated walk complexity $T(x)$ is the minimum length of a walk w, such that the NFA generated from w accepts x uniquely.

For example, $E_N(\varepsilon) = T(\varepsilon) = 0 < 1 = A^{\mathrm{nb}}(\varepsilon)$. For a stronger separation, $E_N(0^{n-1}1) = T(0^{n-1}1) = 2 < n = A^{\mathrm{nb}}(0^{n-1}1)$ for $n \geq 3$.

Theorem 3.23. *We have $E_N \neq T$; in fact, $E^{\pm} \ngeq T$.*

Proof. For the word $x = 00100101$, brute force computation yields that there is only one witness to $E_N(x) \leq 4$, namely, the state sequence 012012013:

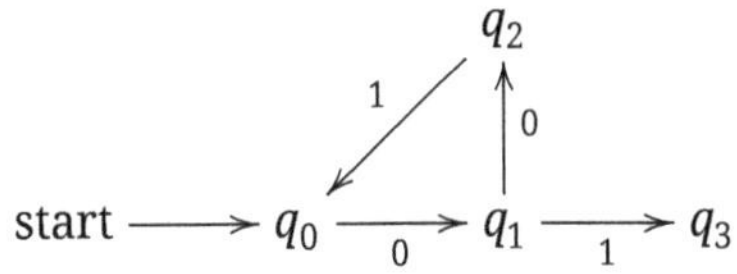

(3.9)

Since the edge (q_0, q_1) must be included twice in any edge-covering walk, this NFA is not a witness to $T(x) \leq 4$ and therefore $T(x) \geq 5$. Since the NFA in (3.9) is bi-deterministic, we also have $E^{\pm}(x) = 4$. □

Definition 3.24. A slight variation $T' \geq T$ requires that the walk end in the final state, as defined by the unique accepting walk for x.

Theorem 3.25. *We have $A^{\circ} \ngeq T'$.*

Proof. Let $x = 010$. Then $A^{\circ}(x) = 2$, but $T'(x) = 3$. (In this case, $A^{\mathrm{nb}}(x) = 2$ and $T(x) = 2$.) □

Remark 3.26. The existence of a minimal-length truncated walk implies that the regex corresponding to the NFA has no $(u+v)^*$, since otherwise we could visit the edges corresponding to u or v in either order. This "truncated walk complexity" counts edges used to glue together two cycles twice, including gluing with the "cycle at infinity" where we connect the start state and the final state with an imagined edge to form a cycle. The truncated walk is obviously always itself a square-free word. As such, it may be a rather compact encoding of the NFA. Of course, this compactness or size, if we were to save the truncated walk in a file, depends on the number of states as well.

There is a corresponding fact about solutions to unconstrained systems of linear Diophantine equations (but see Lemma 3.29 for the constrained case). If we choose n minimal such that $ax+by = n$, say, has a solution with all variables nonzero (necessarily, then, $n = a + b$ and the solution is $(x, y) = (1, 1)$) and for some $m \geq n$ there is a unique solution to $ax + by = m$, then the solution to $ax + by = n$ is unique. For example, $2x + 3y = 7$ has the unique solution $(x, y) = (2, 1)$ and therefore the solution to $2x + 3y = 5$, $(x, y) = (1, 1)$, is also unique. But $2x + 4y = m$ never has a unique totally nonzero solution because we can always make $y = 0$ for another solution.

The computational complexity of verifying that an M is a witness for $A_N(x)$ in terms of an associated Diophantine equation is addressed in Chapter 7. See Exercise 3.3 and Exercise 3.4 for some relevant ideas when the number of variables is two or three.

Diophantine equations have the following *reduction property*, which does not extend in certain ways (see Lemma 3.28 and Lemma 3.29).

Lemma 3.27. *Suppose the equation*

$$ax + by + cz = n \tag{3.10}$$

has a unique solution over $\mathbb{N}$. *Assume that the unique solution* (x_0, y_0, z_0) *satisfies* $x_0 > 0$. *Then the equation*

$$ax + by + cz = n - a \tag{3.11}$$

also has a unique solution.

Proof. Equation (3.11) has at least one solution, namely, $(x_0 - 1, y_0, z_0)$. Next, if (x_i, y_i, z_i) for $i = 1, 2$ are solutions of (3.11), then $(x_i + 1, y_i, z_i)$ are solutions of (3.10), and hence by the injectivity of $x \mapsto x + 1$, we are done. □

See Exercise 3.2 for another version of Lemma 3.27. In addition, there is a multiplicative amplification property (see Exercise 3.1).

Exercise 3.1 does not work in the other direction. For example, $2x + 3y = 7$ has a unique solution $(2, 1)$, but this does not imply that $x + 3y = 7$ has a unique solution.

Lemma 3.27 also does not have a converse: for example, $2x + 3y = 7$ has the unique solution $(2, 1)$, but $2x + 3y = 9$ has, in addition to $(3, 1)$, the solution $(0, 3)$.

Exercise 3.1 has an application to the significance levels for depth of witnesses for A_N (see (7.12)). The length-23 word $x = (01234)^2 5 (678)^4$ has $A_N(x) = 8$ using the fact that the Diophantine equation $5x + 3y = 22$ has the unique solution $(2, 4)$. If we now look at witnesses for the relaxed inequality $A_N(x) \leq 10$, they all involve Diophantine equations obtained from the reduction property; for $A_N(x) \leq 11$ they also arise from the amplification property, namely, $5x + 6y = 22$, via the state sequence 0123401234056789A56789A5.

The reduction property in Lemma 3.27 gives a whole distributive lattice of equations with unique solutions:

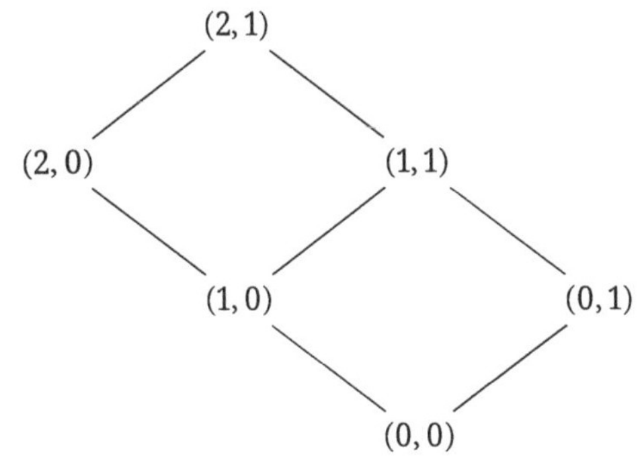

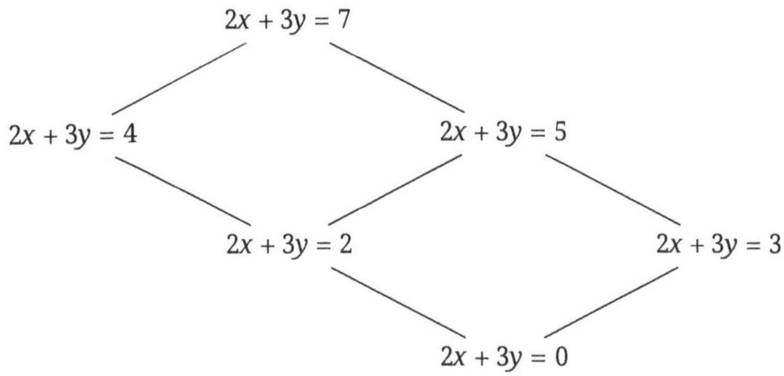

Lemma 3.28. *The following statement is false in general: suppose the equation $\sum_{i=1}^{k} a_i x_i = m$, with some constraints of the form $x_i > 0 \implies x_j > 0$, has a unique solution in $\mathbb{N}^k$. Then the same equation, but without the constraints, has a unique solution as well.*

Proof. A counterexample is $x + 2y = 3$ with the constraint $x > 0 \implies y > 0$. It has the unique solution $(1, 1)$, but without the constraints, another solution is $(3, 0)$. □

Lemma 3.29 shows that the walk witnessing the truncated complexity $T(x)$ will in general not itself be witnessing $A_N(y)$ for any y.

Lemma 3.29. *The following statement is false in general: suppose the equation $\sum_{i=1}^{k} a_i x_i = m$, with some constraints of the form $x_i > 0 \implies x_j > 0$, has a unique solution in $\mathbb{N}^k$. Suppose moreover that the unique solution satisfies $x_i \geq 1$ for all i, and $x_i = 1$ for nonleaves in the constraint structure. Then the equation $\sum_{i=1}^{k} a_i x_i = \sum_{i=1}^{k} a_i$, with the same constraints, has a unique solution as well.*

Proof. Suppose $\sum_{i=1}^{k} a_i u_i = \sum_{i=1}^{k} a_i$ is another solution. Then $\sum_{i=1}^{k}(u_i + x_i - 1) = m$, so we have nonuniqueness for the original equation. However, the constraints may not be satisfied.

The following example was found from an automatic complexity witness for the word $0^6 10^3$, namely, the witnessing sequence of states 01234442301:

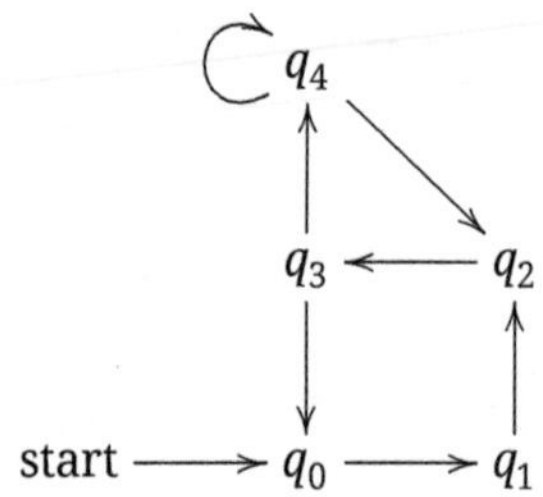

The word has length 10; removing the backbone it has length 9. The example is the system $x + 3y + 4z = 9$ with the constraint $x > 0 \implies y > 0, y > 0 \implies z > 0$. It has the unique solution $(2, 1, 1)$. If we now consider the solution $(1, 1, 1)$ to $x + 3y + 4z = 8$ with the same constraint, it is no longer unique, as a second solution is $(0, 0, 2)$. □

3.5 Covering, plurality, majority

Definition 3.30. Let w be a word. The *covering complexity* $A_{Nc}(w)$ is the minimum number of states of an NFA that accepts w and accepts on only one walk of length $|w|$ that visits every edge.

In covering complexity witnesses there may be several accepting walks of length $|w|$ but only one of them uses every edge. We have $A_{Nc} \le A_N = A_{Nu}$, but it is not immediately clear whether A_{Nc} and A_{Ne} are comparable.

Proposition 3.31. *We have* $A_N \ne A_{Nc}$.

Proof. Let $w = 0110$. We have $A_N(w) = 3$. However, $A_{Nc}(0110) = 2$, using a Kayleigh graph for 010. □

The complexity $A_{Np}(w)$, plurality complexity, requires that there be strictly more accepting walks for the word w than for any other particular word of the same length. We clearly have $A_{Np} \le A_{Ne} \le A_{Nu}$, and we can actually separate two of these.

Proposition 3.32. *We have* $A_{Np} \ne A_{Ne}$.

Proof. Let $w = 10^6 110$. We conducted separate *brute force* verifications of the facts that $A_{Nu}(w) = 5$, $A_{Np}(w) = 4$, and for all words w of length at most 10, $A_{Nu}(w) = A_{Ne}(w)$ (Theorem 3.33). To verify that $A_{Np}(w) \le 4$ the reader may consider the NFA in Figure 3.1. □

Theorem 3.33 (Verified using a computer). *There is no binary word* x *with* $|x| \le 10$ *and* $A_N^{\omega}(x) \ne A_N^{\pi}(x)$.

There are at least four ways to define complexity for NFAs that all coincide for DFAs, and thus could each arguably be the correct “nondeterministic automatic complexity”:

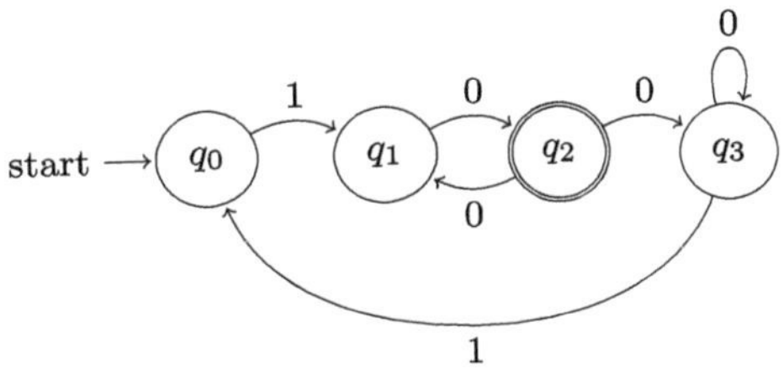

Figure 3.1: A witness to $A_{Np}(10^6110) \leq 4$.

A_{Nu}, A_{Ne}, A_{Np}, and $A^\dagger$. This is because for functions $f : X \to \{0,1\}$ on a set X, but not for functions $f : X \to \mathbb{N}$, the following are equivalent for a fixed $x \in X$:

- $f(x) = 1$ and $f(y) = 0$ for all $y \neq x$ [A_{Nu}];
- $f(y) > 0$ iff $y = x$ [A_{Ne}];
- $f(x) > f(y)$ for all $y \neq x$ [A_{Np}];
- $f(y) = 1$ iff $x = y$ [$A^\dagger$];
- $f(x) > \sum_{y \neq x} f(y)$ [A_{Nm}].

Now, let $f(x)$ be the number of accepting walks of the word x in a given NFA.

Definition 3.34. The *majority complexity* A_{Nm} is defined analogously to A_{Nu} and A_{Ne} using the last condition above, $f(x) > \sum_{y \neq x} f(y)$. Similarly, the *plurality complexity* A_{Np} is defined using the third condition, $f(x) > f(y)$ for all $y \neq x$.

In Proposition 3.32, the counts are $f(1000000000) = 1$, $f(1000000110) = 3$, $f(1000011000) = 2$, $f(1000011000) = 1$, $f(1001100110) = 1$, so that we do not have a majority.

To reduce the computational complexity of $A_{Nm}(x)$ we may hope that rather than searching through all NFAs, it suffices to consider those generated by a single walk of length $|x|$. This is not so clear, as (3.12) shows. Here, there are two accepting walks for $0^4 10^3$ and one for 0^8, and hence we have a witness for $A_{Nm}(0^4 10^3) \leq 6$; nevertheless, each walk generates a sub-NFA that fails to have this property.

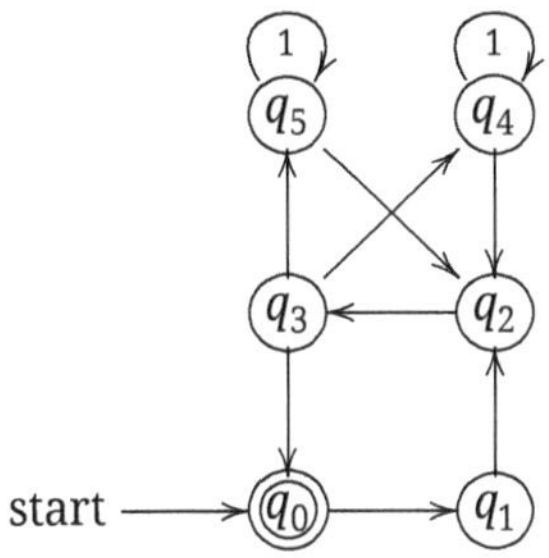

(3.12)

Theorem 3.35 is based on this simple “duplicating states” strategy.

Theorem 3.35. *We have $A_{Nm} \neq A_{Nu}$.*

Proof. Let x be the word $11(010)^5001$, having length 20. We have $A_N(x) = 7$, but $A_{Nm}(x) \leq 6$, as witnessed by (3.13):

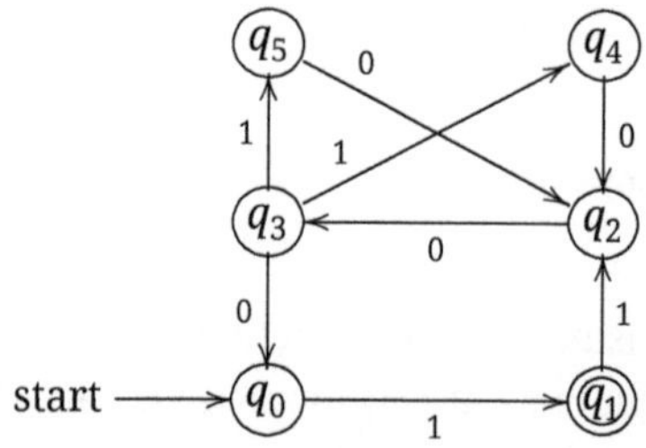

(3.13)

It accepts $11(010)^5001$ in $32 = 2^5$ ways and other words of the same length in eight ways. Indeed, note that the first three and last two moves are forced, and the remaining 15 moves consist of a sequence of four-cycles and three-cycles:

$$(3_1, 4, 4, 4), (3_2, 4, 4, 4), \ldots, (4, 4, 4, 3_2).$$

There are four choices of when to do a three-cycle. For each such choice, there are two choices for which three-cycle to use (3_1 or 3_2). A witness to $A_N(x) \leq 7$ is shown in (3.14):

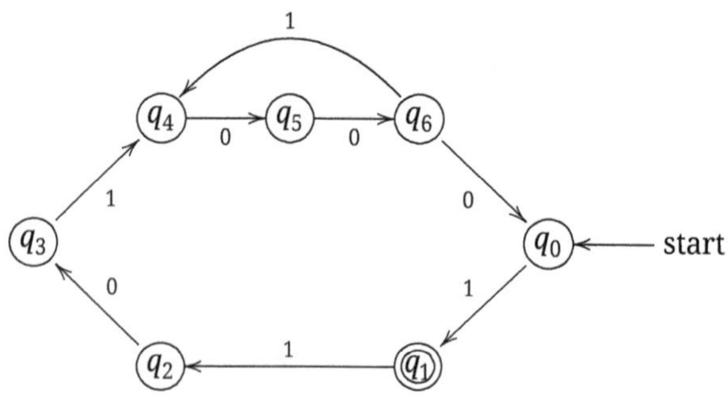

(3.14)

□

Remark 3.36. The NFA used in Theorem 3.35 stretches the notion of "automatic complexity witness" in that it is (crucially) not even a minimal NFA for its language. In fact, our interest in Question 1.3 comes from the fact that A_{Ne} retains Shallit and Wang's property of depending on the automaton M only via the language $L(M)$. We can make things worse: digraphs with multiple equivalent edges (labeled multidigraphs), thus going beyond NFAs:

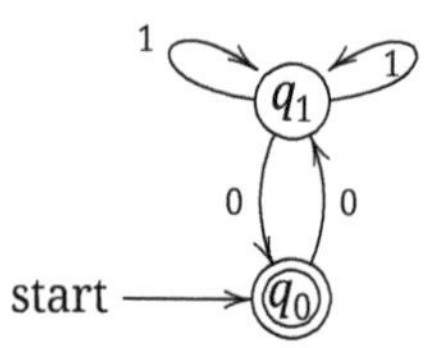

(3.15)

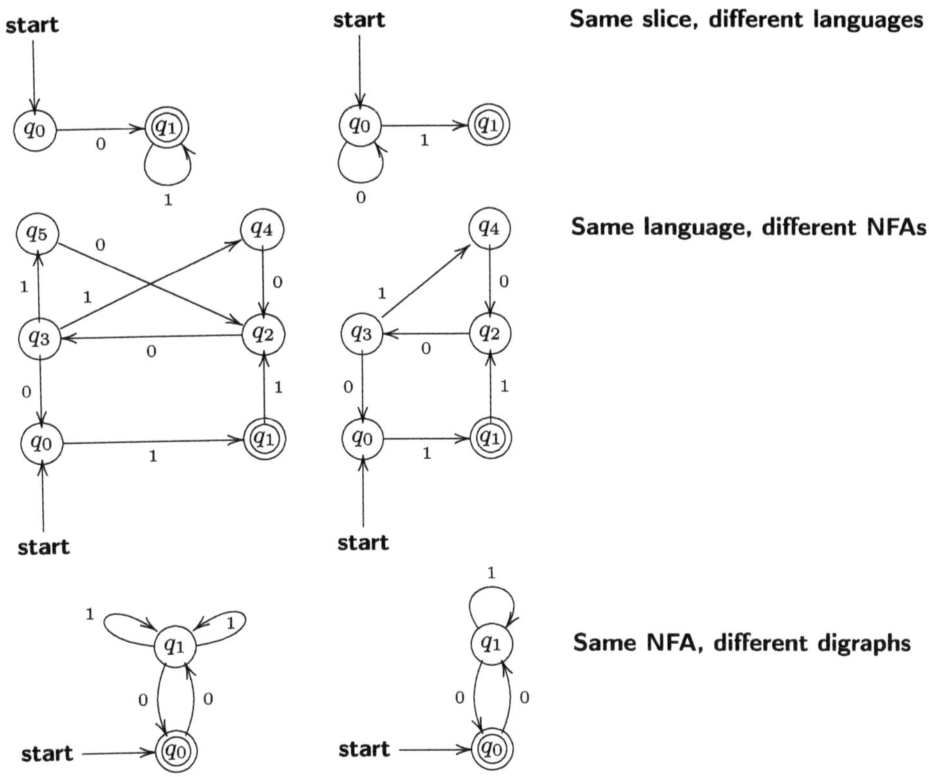

Figure 3.2: Language slices ($L(M) \cap \Sigma^n$, $n = 2$ shown), languages, NFAs, and digraphs.

In (3.15), words of the form 01^*0 have the most accepting walks. For instance, at length $n = 4$, 0110 has $4 = 2^2$ accepting walks, and 0000 has one. Thus, in this *multidigraph complexity*, 0110 has complexity at most 2. See Figure 3.2 for the big picture.

3.6 Bounding deterministic complexity

Theorem 3.37 shows that deterministic (A^-, Definition 1.19) and nondeterministic ($A_N = A_N^\pi$) automatic complexity can differ somewhat: there is a word x such that $A^-(x) - A_N(x) = 2$.

Theorem 3.37. *There exists a sequence x with $A^-(x) - A_N(x) \geq 2$. In fact, there is an m-sequence x with $A^-(x) - A_N(x) = 2$.*

Proof. Let $x = 0001010110100001100100111110111$. A computer run (discussed in [39]) showed that $A^-(x) = 18$ and $A_N(x) = 16$. □

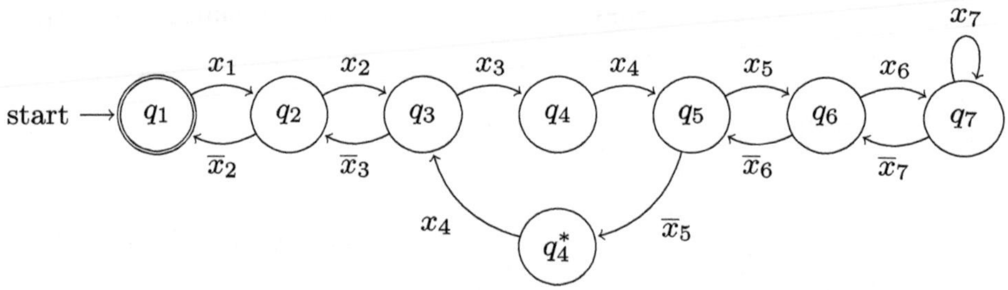

Figure 3.3: A deterministic finite automaton that only accepts one word $x = x_1x_2x_3x_4x_5x_6x_7\overline{x_7x_6x_5}x_4\overline{x_3x_2}$ of length $n = 13$. It is obtained by vertex-splitting the state q_4 in a Kayleigh graph (Figure 2.1).

We show in Theorem 3.38 that for "most" words, the difference between A^- and A_N is small. The case of all words is treated in Question 4.4.

Let $[b] = \{0, \ldots, b-1\}$ and let $|X|$ denote the cardinality of a set X. We show that most words have A^--complexity at most $(\frac{1}{2} + \varepsilon)n$ in the following sense.

Theorem 3.38. *For each $\varepsilon > 0$ and integer $b \geq 1$,*

$$\lim_{n\to\infty} b^{-n}\left|\left\{x \in [b]^n : \frac{A^-(x)}{n} \leq \frac{1}{2} + \frac{1}{2b} + \varepsilon\right\}\right| = 1.$$

Proof. Let $n = 2q - 1$ be odd, $q \geq 1$; the even case is similar. Let M_0 be the Kayleigh graph (Figure 2.1) for x, and let M_1 be the *determinization* of M_0, defined as follows. For each state of M_0 with nondeterministic out-behavior, split it into two states of M_1 as in Figure 3.3. Let $\mathcal{Q}(x)$ be the number of states in the determinization of x. Since the start state is always deterministic, $\mathcal{Q} = q + \sum_{i=1}^{q-1} X_i$, where X_i are independent Bernoulli random variables with parameter $p = 1/b$. The expected number of states of M_1 is therefore

$$\mathbb{E}(\mathcal{Q}) = q + \frac{q-1}{b}.$$

Asymptotically,

$$\mathbb{E}\left(\frac{\mathcal{Q}(x)}{n}\right) \leq \frac{q}{n} + \frac{q-1}{bn} = \frac{n+1}{2n} + \frac{n-1}{2bn} = \frac{1}{2} + \frac{1}{2b} + o(1).$$

By the weak law of large numbers,

$$\lim_{n\to\infty} b^{-n}\left|\left\{x \in [b]^n : \frac{\mathcal{Q}(x)}{n} \leq \frac{1}{2} + \frac{1}{2b} + \varepsilon\right\}\right| = 1.$$

Since $A^-(x) \leq \mathcal{Q}(x)$ for all x, the statement of the theorem follows. □

Theorem 3.39 is a new unconditional upper bound, better than the logarithmic one obtained by Shallit and Wang (Theorem 1.55).

Theorem 3.39. *For all words* x, $A^-(x) \leq |x| + 1 - \sqrt{|x|/2}$.

Proof. By brute force search we can verify that the result holds for words of length at most 7. So assume $|x| \geq 8$. Let $c = |\{k : x_k \neq x_{k+2}\}|$. Thus, $x = a_0 a_1 \ldots a_c$, where each a_i is in $(01)^*, (10)^*, 0^*$, or 1^*. Let $t_i = |a_i|$. Let α be a positive number to be determined below.

Case 1: $c + 1 \leq \alpha\sqrt{n}$. The average of the numbers $t_0, \ldots, t_c$ is

$$\frac{1}{c+1}\sum_{i=0}^{c} t_i = n/(c+1) \geq \frac{n}{\alpha\sqrt{n}}.$$

Thus, there is some i with $t_i \geq \frac{1}{\alpha}\sqrt{n}$.

Case 1.1: $0 < i < c$.

Case 1.1.1: $t_i < n/2$ (roughly). Form a large cycle with a single cycle of length 1 or 2 for a_i attached, to obtain $A^-(x) \leq |x| - (t_i - 1) \leq |x| + 1 - \frac{1}{\alpha}\sqrt{|x|}$.

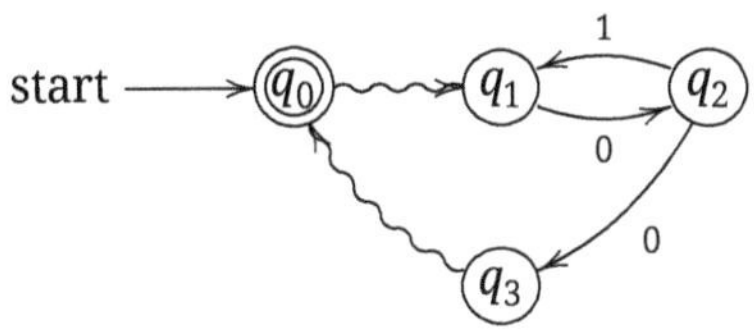

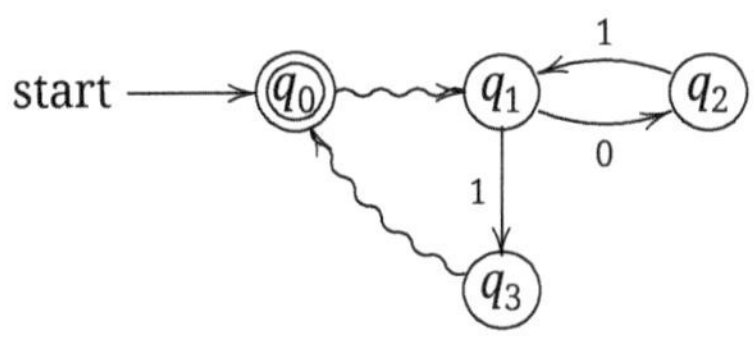

Case 1.1.2: $t_i \geq n/2$. We obtain $A^-(x) \leq n/2$ (roughly).

Case 1.2: $i = 0$ or $i = c$. Attach a small cycle but do not close the large cycle, giving $A^-(x) \leq |x| + 2 - \frac{1}{\alpha}\sqrt{|x|}$.

Case 2: $c + 1 > \alpha\sqrt{n}$. Then we use determinized Kayleigh graphs as in Theorem 3.38, since $x_k \neq x_{k+2}$ implies $x_k \neq x_{n-k}$ or $x_{k+2} \neq x_{n-k}$. If n is odd, we use M_1 and M_2, and if n is even, we use M_1' and M_2'. The final state of M_1 is q_0 and that of M_1' is q_1. The final state of M_2 is q_{u+1} and that of M_2' is q_0.

M_1, M_1'

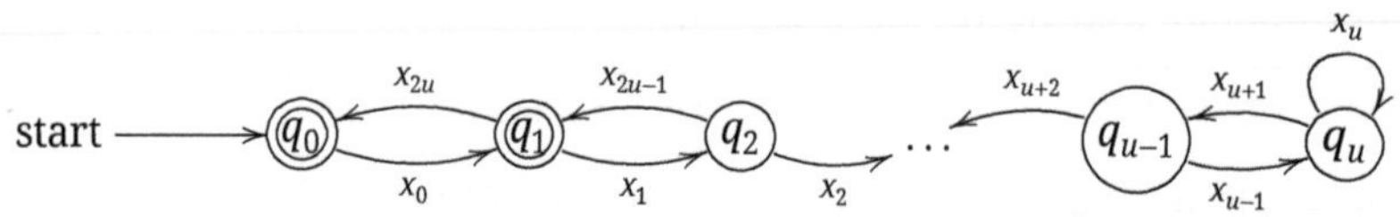

M_2, M_2'

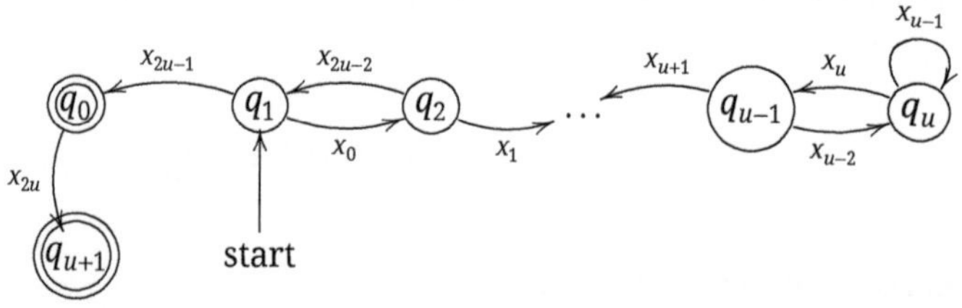

Determinism of M_1 requires $x_{k+2} \neq x_{2u-1-k}$ for each $-1 \le k \le u-2$. Determinism of M_2 requires $x_k \neq x_{2u-1-k}$ for $0 \le k \le u-1$.

Whenever $x_k \neq x_{k+2}$ (which happens for at least $\alpha\sqrt{n}-1$ many k), we thus have either $x_k \neq x_{2u-1-k}$ or $x_{k+2} \neq x_{2u-1-k}$. Thus, there are at least $(\alpha\sqrt{n}-1)/2$ many states in either M_1 or M_2 that are already deterministic.

So

$$A^-(x) \le |x| + 1 - \frac{\alpha\sqrt{|x|}-1}{2}.$$

To get a bound that works in general we now pick α so that $\alpha/2 = 1/\alpha$, namely, $\alpha = \sqrt{2}$. □

Remark 3.40. A worked, random example of Theorem 3.39: Let $n = 17$ and $x = 11100000110010111$. If we only include the case where everything is defined in c, then $c = 9$. Let us checkmark (✓) whenever $x_i = x_{i+2}$.

The result is shown in Table 3.1 with M_1 and M_2. When $x_k \neq x_{k+2}$, either:

(i) the x_{k+2} edge is part of determinism in M_1, or
(ii) the x_k edge is part of determinism in M_2.

The number of blocks of the form 0^*, 1^*, $(01)^*$, or $(10)^*$ is one plus the number of ✗ that "count," where reading from left to right an ✗ counts (indicated by "c") by default, but *does not count* if it is immediately preceded by an ✗ that *does count*. The number c overestimates how many times we have to switch to a different element of $\{(01)^*, (01)^*, 0^*, 1^*\}$. In Case 1 this is good. In this case the best factor is 0^5. We see that six of the nine states in M_1, $q_1, q_2, q_4, q_6, q_7, q_8$, are in need of determinization. Are there $\sqrt{n}-1$ states that are already deterministic? No, not for M_1. For M_2 only q_1, q_2, q_5, q_6 are in need of determinization and $4 = |\{q_3, q_4, q_7, q_8\}| \ge \sqrt{17}-1$ are already deterministic.

Definition 3.41. For an infinite word x, let $I(x)$ denote the infimum of the automatic complexity rate $A(x \upharpoonright n)/n$ as $n \to \infty$.

Table 3.1: Illustration for Remark 3.40.

1	1	1	0	0	0	0	0	1	1	0	0	1	0	1	1	1
✓	✗	✗	✓	✓	✓	✗	✗	✗	✗	✗	✓	✓	✗	✓		
	c					c		c		c			c			
	(i)	(ii)				(ii)	(ii)	(ii)	(ii)	(i)			(ii)			

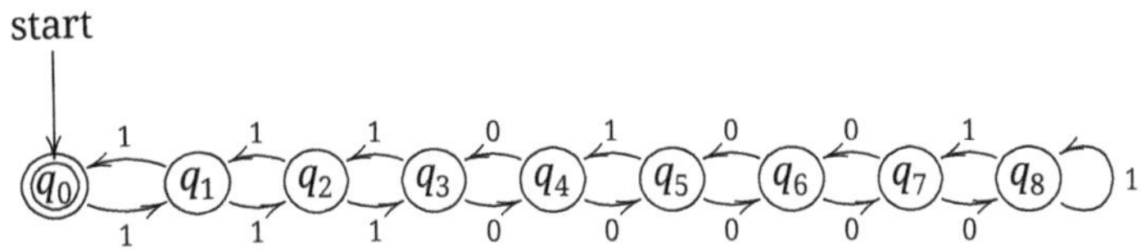

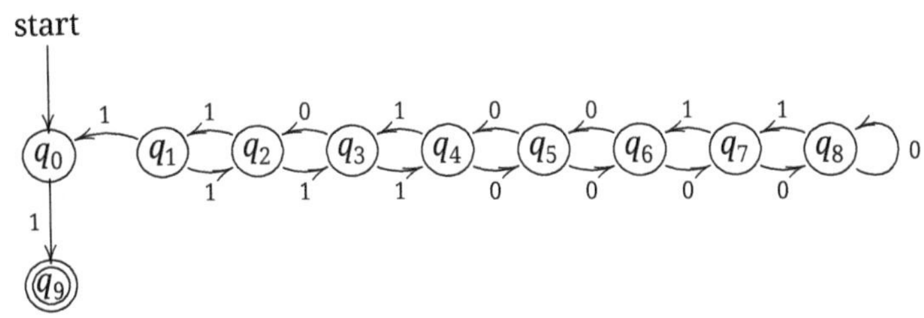

The *Foundations of Software Technology and Theoretical Computer Science* (FSTTCS) conference was held online on December 15–17, 2021. During Liam Jordon's talk at the conference, he said that he and Philippe Moser, when working towards their paper [35], had started by asking themselves whether $I(x) = 1$ necessarily for normal words x. They showed that counterexamples exist but here we turn a $\exists$ to a $\forall$.

Theorem 3.42. *If x is binary and normal to base 2, then $I(x) \leq 3/4$.*

Proof sketch. If x is normal, then 00 and 11 occur no more often than 01 and 10. Hence, if Pr denotes the uniform distribution on $\{1, \dots, n\}$ and we fix x with $|x| = n$, then $\Pr(x_k = x_{k+1}) \leq 1/2 + \varepsilon$. Therefore, $\Pr(x_k = x_{n-k} = x_{k+1}) \leq 1/2 + \varepsilon$. So $\max\{\Pr(x_k \neq x_{n-k}), \Pr(x_{k+1} \neq x_{n-k})\} \geq 1/4 - \varepsilon/2$. Therefore, a Kayleigh graph, or one shifted by one, works deterministically for a large proportion of k and therefore $A(x) \leq 3n/4$. The proof should closely follow that of Theorem 3.39. □

3.6.1 Deterministic shortness

To prove that A^- satisfies the second subword inequality (shortness) is harder than proving this for A_N.

For the transition function δ', if $\delta(q_f, 0)$ and $\delta(q_f, 1)$ are both defined, then we argue that in the accepting path for x, we visit q_f three times: using the 1-edge, using the 0-edge, and at the very end accepting x. But then we have two words $0u$ and $1v$ with $0u1v = 1v0u$, a contradiction.

If only one of $\delta(q_f, 0)$ and $\delta(q_f, 1)$ is defined and we are lucky with the first symbol of z, then we are also okay. A problematic case is where y has an automaton with two cycles with a certain unique solution to a corresponding Diophantine equation and z consists in continuing the second cycle for a while. Then continuing it may well break the unique solvability.

We can demonstrate that there exists a word $w :; a$ where x has an A^- witness which does not extend (here we allow for the addition of edges and a state and for changing the final state only) to an A^- witness for $w :; a$ with at most one more state. Namely, for $x = 0^6 1^6$, the state sequence 0123123454545, we get the equation $3x + 2y = 7$, and if we extend with $a = 1$ we get $3x + 2y = 9$, which has the multiple solutions $(3, 0)$ and $(1, 3)$.

Instead, to prove Theorem 3.44 we will use vertex splitting again. Unlike in Theorem 3.38, however, we will use only a partial vertex splitting where some edge behavior is shared between the splits.

Definition 3.43. Let $G = (V, E)$ be a labeled digraph with vertex set V and labeled edge set E. A *partial vertex splitting of G at $v \in V$* is a digraph $G' = (V', E')$, where $V' = V \cup \{v'\}$, with the following properties. Let $v_1, v_2 \in V$.

1. When restricted to edges not incident to v or v', E' is equal to E.
2. If $(v_1, v, b) \in E'$ or $(v_1, v', b) \in E'$ (an edge from v_1 to v with label b), then $(v_1, v, b) \in E$.
3. If $(v, v_2, b) \in E'$ or $(v', v_2, b) \in E'$, then $(v, v_2, b) \in E$.

In the case that we never have both $(v_1, v, b) \in E'$ and $(v_1, v', b) \in E'$ and never have both $(v, v_2, b) \in E'$ and $(v', v_2, b) \in E'$, the partial vertex splitting is simply a *vertex splitting*.

Theorem 3.44. *The deterministic complexity functions A^- and A both satisfy shortness: $A^-(x :; a) \leq A^-(x) + 1$ and $A(x :; a) \leq A(x) + 1$.*

Proof. Let q_f be the final state of a witness for $A^-(x)$. If $\delta(q_f, a)$ is undefined (empty), then a very simple construction with a new state and a new edge is carried out. If $\delta(q_f, a)$ is defined (i. e., nonempty), then we use a partial vertex splitting at the state $\delta(q_f, a)$.

Construction of M'. Given $M = (Q, \Sigma, \delta, q_0, q_f)$, x, a we modify it as follows.

If there is no outgoing edge from q_f labeled a, let $q' \notin Q$, let δ' extend δ, the difference being only that $\delta'(q_f, a) = q'$, and let $M' = (Q \cup \{q'\}, \Sigma, \delta', q_0, q')$.

Otherwise, as in Figure 3.4, let $q = \delta(q_f, a)$, let $q' \notin Q$, and let $M' = (Q', \Sigma, \delta', q_0, \delta(q_f, a))$; for all $r \in Q \setminus \{q_f\}$, if $\delta(r, b) = q$, then let $\delta'(r, b) = q'$ instead. Also, whenever $\delta(q, b) = s$, then we let $\delta'(q', b) = s$. Moreover, we remove any relationship of the form $\delta(r, b) = q$ except for the case $(r, b) = (q_f, a)$. This has the effect of making the states q and q' be like q but now with memory of whether they were reached by reading a from q_f

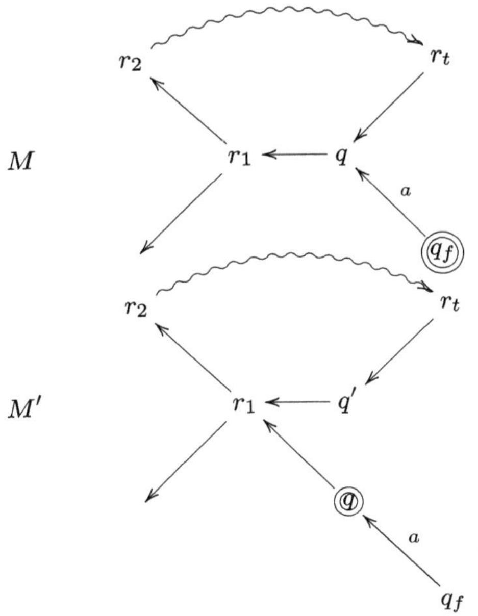

Figure 3.4: The hardest case of the second subword inequality for A^-.

or not. The construction does not create any new walks to the final state because walks entering q and q' must have come from distinct states. □

As an example of the shortness construction in Theorem 3.44, a witness for the A^--complexity of the nonrepeating part of the Fibonacci numbers mod $p = 8$, the length-12 word

$$x = (0, 1, 1, 2, 3, 5, 0, 5, 5, 2, 7, 1)$$

is shown in (3.16), together with the result of applying the shortness construction to it, to get a witness for $A^-(x :; 1) \le 8$:

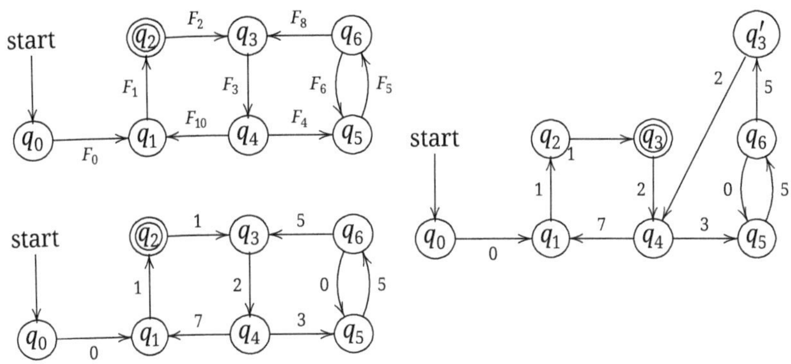

(3.16)

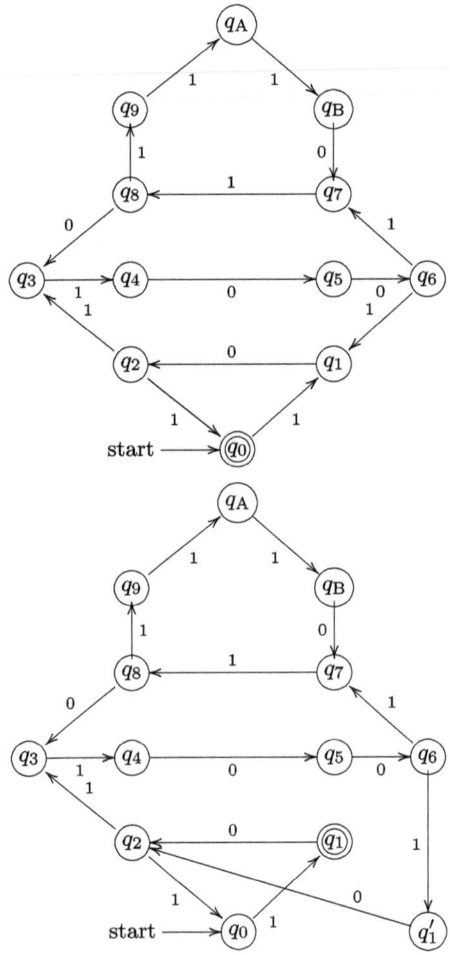

Figure 3.5: Above: a witness M for $A_N(x) \le 12$ for $x = 1011001111101111010100101$. Below: M', which is M modified according to the A^- shortness construction (Theorem 3.44) passing from x to $x1$. Since the new final state q_1 has in-degree greater than one, a new state q_1' is created. Incidentally, neither M nor M' is deterministic. Serendipitously, this M' turns out to be an optimal witness for $x1$, showing that $A_N(x1) \le 13$. It is one of six witnesses that have 13 states and 16 edges.

Another example of the shortness construction in Theorem 3.44 is given in Figure 3.5.

Remark 3.45. We can also give an example that demonstrates that sometimes, our shortness construction in Theorem 3.44 gives the only possible automaton and in that sense is the only game in town. Let $x = 0010100$. The only witness for $A^-(x) = 3$ is the state sequence 01212120:

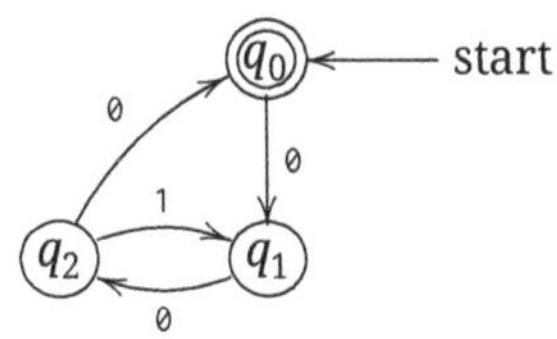

(3.17)

In fact, this is even the only witness for $A_N(x) = 3$, as mentioned in Example 1.66. The only witness for $A^-(x :; 0) = 4$ is the state sequence 012323201:

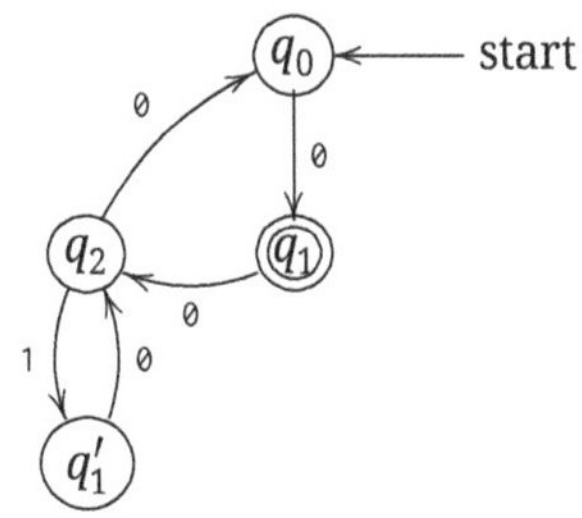

(3.18)

which is exactly what our shortness construction gives.

Here the state 3 or q_1' is the split version of the state 1 or q_1. There is no shorter word than 0010100 with this property, even in larger alphabets. Moreover, 0010100 is the only word over $\{0,1\}$, starting with 0, of its length with this property, although there are a couple of other examples over larger alphabets of the same length.

3.7 Open problems

Question 3.1. Concerning the "profiles" $(A_N(x), E_N(x))$ and $(A^-(x), E^-(x))$, at length 7, the word 0001101 can be accepted with five states and five edges deterministically and with four states and no less than six edges nondeterministically. We have $A_N \le E_N$, $A^- \le E^-$ for nonempty words. Can we demonstrate that A^- and E_N are incomparable? The example just mentioned shows that we can have $A^-(x) = 4 < 5 = E_N(x)$. To get $E_N(x) < A^-(x)$ we would need a stronger example than that of Theorem 3.11.

Question 3.2. Is the following problem efficiently solvable?
- Instance: a word x.
- Question: is $E_N(x) = |x| - 1$?

3.8 Exercises

Exercise 3.1. Prove a version of the amplification of $ax + by = n$ mentioned above: if the unique solution of $ax + by = n$ is (x_0, y_0) with $y_0 = uv_0$, where $u \neq 0$, then the unique solution of $ax + buv = n$ is $(x, v) = (x_0, v_0)$.

A complete solution to Exercise 3.1 is available at the following URL.

https://github.com/bjoernkjoshanssen/ac-exercises/

Exercise 3.2. Prove a version of the reduction in Lemma 3.27 for the two-variable equation $ax + by = n$.

A complete solution to Exercise 3.2 is available at the following URL.

https://github.com/bjoernkjoshanssen/ac-exercises/

Exercise 3.3. Show that if a and b are relatively prime integers, then there exist integers x, y such that for all integers k, $(u, v) = (x - kb, y + ka)$ satisfies $au + bv = n$.

A complete solution to Exercise 3.3 is available at the following URL.

https://github.com/bjoernkjoshanssen/ac-exercises/

Exercise 3.4. Let $n \in \mathbb{Z}$ and let $a, b, c \in \mathbb{Z}$ be relatively prime as a triple, i. e., $\gcd(\gcd(a, b), c) = 1$. Prove that there exist integers x, y, z such that $ax + by + cz = n$.

A complete solution to Exercise 3.4 is available at the following URL.

https://github.com/bjoernkjoshanssen/ac-exercises/

4 The many variants

4.1 Master diagram

In (4.1), we consider only nonempty words and alphabets Σ with $|\Sigma| \geq 2$. $A \rightarrow B$ means $A \leq B$ as in Definition 4.1. Definitions may be quickly found via the Index at the end of the book.

Definition 4.1. We order functions $C : \mathbb{N} \rightarrow \mathbb{N}$ by $C_1 \leq C_2$ iff $C_1(x) \leq C_2(x)$ for all x, and $C_1 < C_2$ if $C_1 \leq C_2$ but $C_1 \neq C_2$.

Functions known to be efficiently computable are shown in boxes. "Important" functions are shown in ovals.

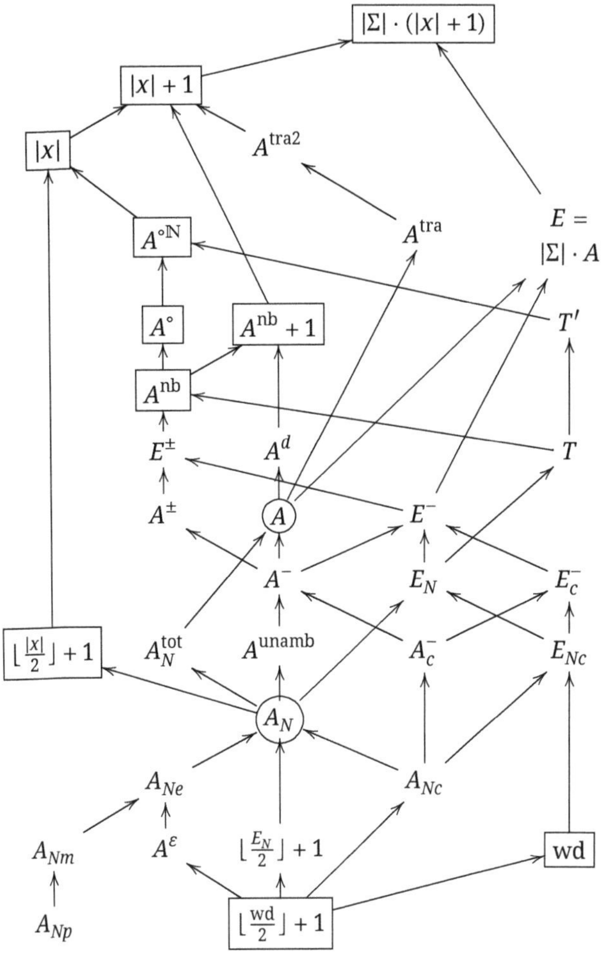

(4.1)

https://doi.org/10.1515/9783110774870-004

The following tables summarize where some implication and nonimplication results are proved.

Function	$\geq$	$>$?	**Witness ($\Sigma = \{0, 1\}$)**
$A^{\text{perm}}(x)$			
	Theorem 4.40	Theorem 4.46	000
$A^{\text{tra}}(x)$			
			00
$A(x)$			
	Theorem 1.20	Theorem 1.20	0
$A^{-}(x)$			
		Theorem 4.30	00011101
$A^{\text{unamb}}(x)$			
		Theorem 4.33	00100001
$A_N(x)$			
		Theorem 4.8	001110010100100001
$A^{\varepsilon}(x)$			

Function	$\geq$	$>$?	**Witness**
$\lvert x \rvert$			
	Lemma 4.20	Theorem 4.19	00
$A^{\circ}(x)$			
	Lemma 4.20	Theorem 4.21	011
$A^{\text{nb}}(x)$			
	Theorem 4.22		001
$E^{\pm}(x)$			
	Theorem 4.22	Theorem 4.28	012, 00110
$A^{\pm}(x)$			
		Theorem 4.36	011100
$A^{-}(x)$			

Map of research

Arrows indicate some cases where papers cite each other, or could have cited each other, without necessarily a strong link.

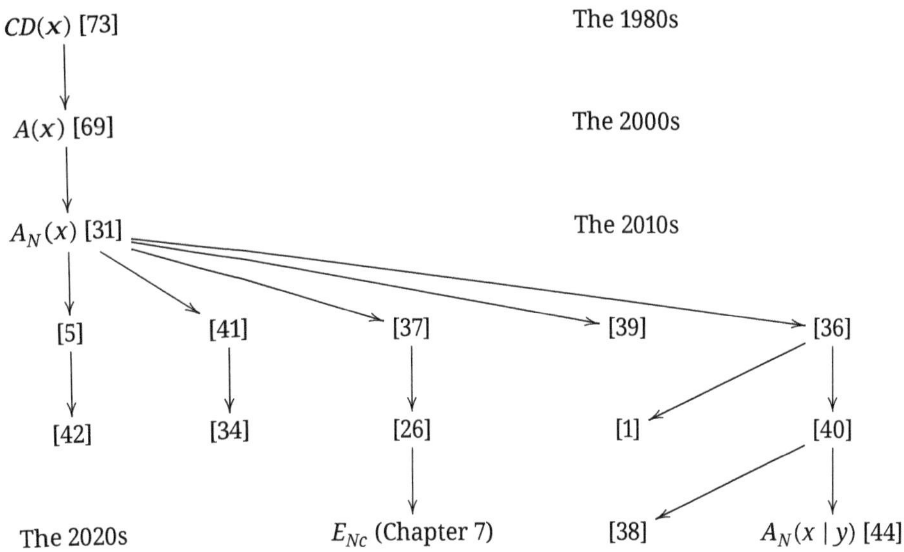

Definitions of proofs for many parts of (4.1) follow.

Definition 4.2. Let A^d denote the version of A where we require that a dead state be present.

Definition 4.3. Let A_N^{tot} denote the version of A_N where we require totality. That is, the transition function δ has the property that for all $q \in Q$ and $a \in \Sigma$, $\delta(q, a)$ is nonempty.

Theorem 4.4. *The total nondeterministic complexity A_N^{tot} is incomparable with both the length function $x \mapsto |x|$ and A^-.*

Proof. We have $A_N^{\mathrm{tot}}(\emptyset) = 2 > 1 = |\emptyset|$. For the other direction, let x be such that $A^-(x) = A^{\mathrm{unamb}}(x) + 2$ (Theorem 4.32). Then

$$A_N^{\mathrm{tot}}(x) \leq A_N(x) + 1 \leq A^{\mathrm{unamb}}(x) + 1 = A^-(x) - 1 < A^-(x).$$

□

Theorem 4.5. *We have $A \neq A^d$. In fact, $A^d \not\leq A^{\mathrm{tra}}$.*

Proof. For example, $A(00010111) = 5$, as witnessed only by a single DFA, which has no dead state:

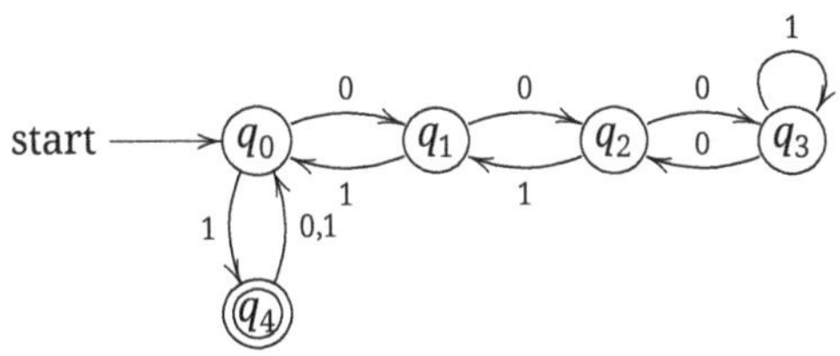

Thus, $A(x) = 5 < A^d(x)$ for $x = 00010111$. Since this example is transitive, we have established $A^d \not\leq A^{\text{tra}}$ in fact. □

Remark 4.6. The example in Theorem 4.5 shows several small things. We might wonder whether simply $A_N^{\text{tot}} = A_N + 1$ and $A = A^- + 1$ for nonempty words, but this is not so. For $x = 00010111$, we have $A(00010111) = 5 = A_N(x) = 5$ so that all four complexity functions agree on x.

Definition 4.7. Let Σ be an alphabet and let $e \notin \Sigma$. The morphism $\pi = \pi_{e,\Sigma}$ is defined by $\pi(a) = a$ for all $a \in \Sigma$, and $\pi(e) = \varepsilon$, the empty word. The ε-complexity of x is defined by $A^\varepsilon(x) = \min\{A_N(y) \mid \pi(y) = x\}$.

Theorem 4.8. *There exists a word x with* $A^\varepsilon(x) < A_N(x)$.

Proof. Let $y = \texttt{1100012101011011110} \in \{\texttt{0},\texttt{1},\texttt{2}\}^{19}$ and $x = \texttt{110001101011011110} \in \{\texttt{0},\texttt{1}\}^{18}$. Let π be the morphism

$$\pi(0) = 0, \quad \pi(1) = 1, \quad \pi(2) = \varepsilon.$$

We have $x = \pi(y)$ (Exercise 4.3). Since $A^\varepsilon(x) \leq A_N(y)$, it suffices to prove that $A_N(y) \leq 9 < 10 \leq A_N(x)$. (As a matter of fact, $A_N(y) = 9 < 10 = A_N(x)$ holds.) We verified the equality $A_N(x) = 10$ in Python on November 28, 2021.

We can prove the fact that $A_N(y) \leq 9$ by studying the 9-state, 11-edge NFA in (4.2):

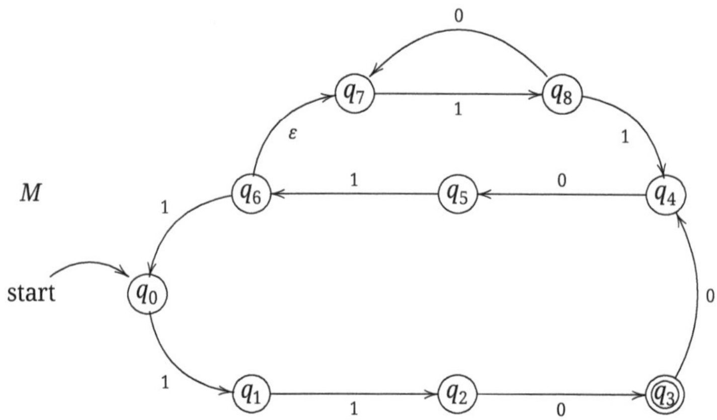

(4.2)

M accepts y uniquely along the state sequence 01234567878784560123, since the problem

$$a, b, c \in \mathbb{N}, \quad 3 + 7a + 5b + 2c = 19,$$
$$c > 0 \to b > 0, \quad b > 0 \to a > 0$$

has the unique solution $(a, b, c) = (1, 1, 2)$. □

4.2 Nonbranching complexity

Definition 4.9. We define three classes of deterministic incomplete DFAs.

1. An NFA is *nonbranching* for each $q \in Q$ and all $a, b \in \Sigma$, if $\delta(q, a) \neq \emptyset \neq \delta(q, b)$, then $a = b$. The set of all such NFAs is denoted by $\mathcal{N}_{\mathrm{nb}}$.
2. The set of all NFAs M such that both M and the digraph-theoretic reverse of M belong to $\mathcal{N}_{\mathrm{nb}}$ is denoted by $\mathcal{N}_{\circ}$.
3. The set of all NFAs $M \in \mathcal{N}_{\circ}$ such that the start state of M equals the final state of M is denoted by $\mathcal{N}_{\circ\mathbb{N}}$.

These classes have corresponding complexity functions:

1. The *non-branching complexity* $A^{\mathrm{nb}}(x)$ is the minimum number of states of an NFA $M \in \mathcal{N}_{\mathrm{nb}}$ such that M uniquely accepts x.
2. The *circular complexity* $A^{\circ}(x)$ is the minimum number of states of an NFA $M \in \mathcal{N}_{\circ}$ such that M uniquely accepts x.
3. The *integral circular complexity* $A^{\circ\mathbb{N}}(x)$ is the minimum number of states of an NFA $M \in \mathcal{N}_{\circ\mathbb{N}}$ such that M uniquely accepts x.

For an alphabet Σ, the set Σ^n for a particular $n \in \mathbb{N}$ is called a *level* (alternatively, *slice*) of Σ^*.

Definition 4.10. A function $C : \Sigma^* \to \mathbb{N}$ is *level-convex* if for each $x, y \in \Sigma^n$ and $k \in \mathbb{N}$ with $C(x) < k < C(y)$, there is $z \in \Sigma^n$ with $C(z) = k$, and *convex* if for each $x, y \in \Sigma^*$ and $k \in \mathbb{N}$ with $C(x) < k < C(y)$, there is $z \in \Sigma^*$ with $C(z) = k$.

Arguably, whether convexity holds is not "natural" since for Kolmogorov complexity, the answer may depend on the choice of universal Turing machine. To get the idea, imagine that a Turing complete programming language can certainly have the property that all programs written in it have even length. Conversely, it seems plausible that there are programming languages where all lengths are those of some optimal programs.

The asymptotic distribution of A^{nb} was studied by Birns [7], who obtained a formula for the number of words of length n of complexity q when $n \geq 2q - 1$. The formula does not depend on n (except for the requirement $n \geq 2q - 1$).

A similar but simpler formula obtains for the number of words of A° complexity q when $n \geq 2(q - 1)$. It is simply the number of primitive words of length q.

For $A^{\circ\mathbb{N}}$ we get a formula that works for all q in Theorem 4.11 below. Let $\mathcal{P}(n, m)$ be the number of primitive words of length n, in an alphabet of size m.

Theorem 4.11. *We have*

$$\#\{x \in \Sigma^n \mid A^{\circ\mathbb{N}}(x) = q\} = \begin{cases} \mathcal{P}(n, m) & \text{if } q \mid n, \\ 0 & \text{otherwise.} \end{cases}$$

Table 4.1: The distributions $\#\{x \in \{0,1\}^* \mid B(x) = q\}$ for $n = 6$ and a binary alphabet, for words starting with 0, for B one of these three complexity functions. An asterisk (*) indicates a value of q for which the asymptotic formulae apply. (For A_N, see Table 2.1.)

q	$A^{\circ\mathbb{N}}$	A°	A^{nb}	A_N
1	1	1*	1*	1*
2	1	1*	2*	3*
3	3	3*	6*	13
4	0	6*	13	15
5	0	11	9	0
6	27	10	1	0
≥ 7	0	0	0	0
level-convex?	✗	✓	✓	?

Corollary 4.12. *The integral circular complexity $A^{\circ\mathbb{N}}$ is not level-convex.*

Proof. Let $n = 3$. We have $A^{\circ\mathbb{N}}(000) = 1 < 3 = A^{\circ\mathbb{N}}(001)$, but since 2 does not divide 3, there is no $x \in \{0,1\}^n$ with $A^{\circ\mathbb{N}}(x) = 2$. □

Theorem 4.13. *The circular complexity A° is level-convex.*

Proof. For each $k \geq 0$, the word 0^k1 is primitive (see Exercise 1.11) and $A^{\circ}((0^k1)^a) = k+1$ for each rational $a \geq 1$. Indeed, if we could write $(0^k1)^a = z^b$ with $|z| < |0^k1|$, then $z = 0^{|z|}$, which is a contradiction. Now choose $a = n/(k+1)$, so that $a(k+1) = n$. □

Theorem 4.14. *The nonbranching complexity A^{nb} is level-convex.*

Proof. For each $1 \leq k \leq n$, let $x_k = 0^{n-k}1^k$. Then $A^{\mathrm{nb}}(x_k) = n+1-k$. □

The situation is summarized in Table 4.1.

Remark 4.15. To prove that A^{-} is convex, i. e., an onto function, we can use words of the form 0^k1^l, where l is large enough that the only optimal witness hard codes the k many zeros.

To prove convexity for $A^{\circ\mathbb{N}}$ we need the result that at each length there exists a primitive word (Exercise 1.11). For A° and A^{nb} we get convexity similarly using Exercise 1.12 and Exercise 1.13.

Theorem 4.16. *The nondeterministic complexity A_N is convex (Definition 4.10).*

Proof. It suffices to show that for each $k \in \mathbb{N}$, there is an x with $A_N(x) = k$. By Theorem 2.1, $\sup\{A_N(x) \mid x \in \Sigma^*\} = \infty$, so let y be such that $A_N(y) > k$. Let $s_i = A_N(y \upharpoonright i)$ for each i. By Theorem 1.47, we can apply Lemma 1.45 and conclude that there is some i with $s_i = A_N(y)$. □

Theorem 4.17. *Let x be a word. There is only one witnessing NFA for $A^{\circ}(x)$.*

Proof. Let $x = y^{a}$ for a minimal-length y. Then a cycle of length $|y|$ is the unique witness for $A^{\circ}(x)$. □

Theorem 4.18. *There exists a word x such that there is more than one witnessing NFA for $A^{\mathrm{nb}}(x)$.*

Proof. We obtain $uv^{a} = xy^{b}$ with $|uv| = |xy|$ but $|u| \neq |x|$:

$$0010 = 0(01)^{3/2} = (001)^{4/3}.$$

□

The trivial Theorem 4.19 provides one sharp separation for Equation (4.1).

Theorem 4.19. *Let $n \geq 1$. For $x = \mathtt{0}^{n}$, we have $A^{\circ\mathbb{N}}(x) = 1 < n = |x|$.*

Proof. The following NFA witnesses the fact that $A^{\circ\mathbb{N}}(x) = 1$.

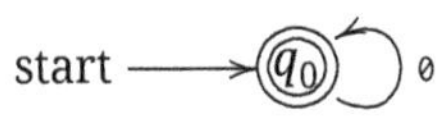

□

Lemma 4.20. *Let x be a nonempty word. Then $A^{\circ}(x)$ is the minimum $|y|$ where $x = y^{a}$, $a \in \mathbb{Q}$. (We can always take $x = y$ and $a = 1$, so $A^{\circ}(x) \leq |x|$.) Similarly, $A^{\mathrm{nb}}(x)$ is the minimum $|yz|$ where $x = yz^{a}$, $a \in \mathbb{Q}$. (We can always take $y = \varepsilon$, so $A^{\mathrm{nb}}(x) \leq A^{\circ}(x)$.)*

Proof. When M and its reverse are both nonbranching, the digraph of the NFA is a path or a cycle. □

Theorem 4.21. *For $x = \mathtt{011}$, we have $A^{\mathrm{nb}}(x) = 2 < A^{\circ}(x) = 3$.*

Proof. Witnesses for $A^{\mathrm{nb}}(x) \leq 2$ and $A^{\circ}(x) \leq 3$:

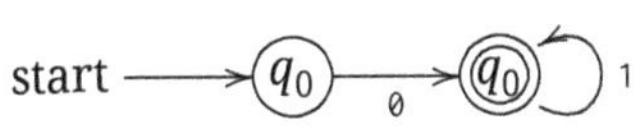

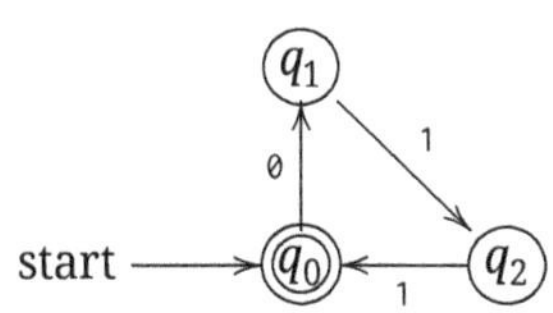

Since 011 is not a power w^{a}, $a > 1$, by Lemma 4.20 $A^{\circ}(x) \geq |x| = 3$. □

Theorem 4.22. *We have $A^{\pm} \leq \max\{E^{\pm}, 1\} \leq A^{\mathrm{nb}}$.*

Proof. Let M be a witnessing NFA for $A^{\mathrm{nb}}(x)$. The reverse of M, while it will generally be branching, will be deterministic, or else one could increase the size of the cycle in M. Moreover, its number of edges equals its number of states. Thus, $E^{\pm} \leq A^{\mathrm{nb}}$.

Now let M' be a witnessing NFA for $E^{\pm}(x)$. Then the number of edges in M' is at least the number of states of M', or else we have just a single path, which we can connect to a single cycle instead. Thus, $A^{\pm}(x) \leq E^{\pm}(x)$ whenever $E^{\pm}(x) \geq 1$. □

Theorem 4.23. *We have* $\lfloor \frac{E_N(x)}{2} \rfloor + 1 \leq A_N(x)$.

Proof. Each state that is not the start state has graph-theoretic degree (deg) at most 4. The start state has degree at most 3. Therefore, considering a witnessing automaton M, with set of states Q, for $A_N(x)$, we have

$$\begin{aligned} E_N(x) &\leq (\text{the number of edges of } M) \\ &= \sum_{v \in Q} \deg(v)/2 \leq \frac{3 + 4(|Q| - 1)}{2} = 2A_N(x) - \frac{1}{2}. \end{aligned}$$

Hence,

$$\begin{aligned} \left\lfloor \frac{E_N(x)}{2} \right\rfloor + 1 &\leq \left\lfloor \frac{2A_N(x) - 1/2}{2} \right\rfloor + 1 \\ &= \left\lfloor A_N(x) - \frac{1}{4} \right\rfloor + 1 \\ &= A_N(x). \end{aligned}$$

□

Definition 4.24. The *width* of x, that is, the number of distinct symbols of x, is denoted $\mathrm{wd}(x) = p_x(1)$.

If we think of a word as a function, the length is the cardinality of the domain and the width is the cardinality of the range. Pictorially, a word like banana has its length and width represented as follows:

b	a	n
√		
	√	
		√
	√	
		√
	√	

The common convention that *width is less or equal to length* becomes the statement that *the cardinality of the range is bounded by the cardinality of the domain.* This is a well-known equivalent of the axiom of choice in general, but a triviality in our (finite) case: we simply map each occurring symbol to its first occurrence.

Width is a special case of *subword complexity.*

Definition 4.25. A word w is a *factor*, or *contiguous subsequence*, of a word v if $v = awb$ for some words a, b. For a finite word x, x^∞ is the infinite word satisfying $x\,x^\infty = x^\infty$. For a finite or infinite word u, $p_u(k)$ is the number of distinct factors of u of length k. The cyclic subword complexity of x is

$$(p_{x^\infty}(1), \ldots, p_{x^\infty}(|x|)).$$

The plain subword complexity of x is

$$(p_x(1), \ldots, p_x(|x|)).$$

For $k = 1$, $p_x(k) = |x|$ is enough to get maximal A_N-complexity. For $k = 2$, a suitable counterexample comes from the theory of power-bordered words: let $n = 6$ and consider the word *abccda*, where a, b, c, d are distinct. Then all subwords of length 2 are distinct, but nevertheless the word does not have maximal A_N-complexity. Now, Theorem 4.26 says that a binary word of maximal subword complexity (among binary words) need not have maximal A_N-complexity.

Theorem 4.26. *There exists a binary word of maximal cyclic subword complexity (for binary words of its length) that does not have maximum A_N-complexity (for words of its length).*

Proof. It is easily checked[1] that already at length 6, we have a string of maximal subword complexity but not maximal A_N-complexity, namely, x = `011100`. It has $(p_x(1), p_x(2), p_x(3), p_x(4), p_x(5), p_x(6)) = (2, 4, 6, 6, 6, 6)$, which is the highest possible for binary words of length 6. However, $A_N(x) = 3$. □

Remark 4.27. As an application of Theorem 4.23 (the application that prompted the result, in fact), consider the first 13 terms of the Collatz sequence [75]

$$0, 1, 7, 2, 5, 8, 16, 3, 19, 6, 14, 9, 9.$$

The *vigesimal* number system has 20 symbols:

0 1 2 3 4 5 6 7 8 9 A B C D E F G H I J.

In it we can express the first 13 terms of the Collatz sequence as

0 1 7 2 5 8 G 3 J 6 E 9 9.

Equivalently, declare that a_n, $n \in \mathbb{N}$, are distinct symbols and consider

1 Apparently not always easily enough, however. In the journal article [39] where this result first appeared, the word `001100` was given, which has $p_x(3) = 5 < |x|$.

$$a_0a_1a_7a_2a_5a_8a_{16}a_3a_{19}a_6a_{14}a_9a_9.$$

Since there are 12 distinct symbols in this sequence of 13 symbols, we have

$$12/2+1=\mathrm{wd}(x)/2+1\le A_N(x)\le 13/2+1,$$

and hence $A_N(x) = 7$. Thus, Theorem 4.23 gives a simple method for proving maximal complexities of some nonsquare-free words.

Theorem 4.28. *There exists a word x with $A^{\pm}(x) < E^{\pm}(x)$.*

Proof. Let $x = $ `012`. Then $A^{\pm}(x) = 2 < 3 = \mathrm{wd}(x) \le E^{\pm}(x)$ using a Kayleigh graph (Figure 2.1). There is no example in a unary alphabet. In a binary alphabet, there is no example of length 4 or less. At length 5 we have the example 00110 with a witness given by the state sequence 012220. □

4.3 Unambiguous complexity

Definition 4.29. An NFA M is *unambiguous* if for each word x, if M accepts x then M accepts x on only one walk. Let w be a word. The *unambiguous complexity* $A^{\mathrm{unamb}}(w)$ is defined identically to $A_N(w)$ except for the added requirement that the witnessing NFAs must be unambiguous.

Clearly, $A_N \le A^{\mathrm{unamb}} \le A^{-}$. We proceed to separate all three.

Theorem 4.30. *We have $A^{\mathrm{unamb}} \ne A^{-}$.*

Proof. Consider the word 00011101. It has an unambiguous NFA witnessing $A_N(x) \le 4$. See Theorem 3.15. The unordered pairs of states $\{q_i, q_j\}$ such that there exists a word w which leads both to q and to q' are $\{q_i\}$ for each i (of course) and $\{q_3, q_1\}$ with $\{q_3\} \to_1 \{q_3, q_1\}$, shown in (4.3):

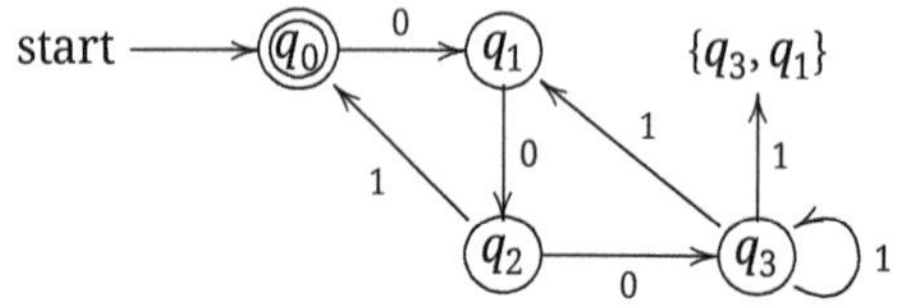

(4.3)

Here the transition $\{q_a, q_b\} \to_i \{q_c, q_d\}$ means that there is a word w that leads to q_a, then to q_c via i, and also from q_b to q_d via i; or the other way around. We use the code in Figure 4.1 to check for ambiguity. There, `t` is a set of tuples `(p1, p2, b)`, where `p1 in range(0,self.q)`, `p2 in range(0,self.q)`, and `b in range(0,self.q)` are true; `(p1,p2,b) in t` means that there is a transition from `p1` to `p2` labeled `b`; and `twoPathsLeadTo` is a set of pairs `(p1, p2)` such that there are two distinct paths, one leading to `p1`, the other to `p2`, with the same word being read. `qFinal` is the final state,

```
def unambiguous(self, t, qFinal):
        twoPathsLeadTo = set()
        for q1 in range(0, self.q):
                for r in range(0, self.q):
                        for s in range(0, self.q):
                                if r != s:
                                        for b in range(0, self.b):
                                                if (q1,r,b) in t and
    (q1,s,b) in t:

    twoPathsLeadTo.update({(r,s),(s,r)})
        while True:
                oldTPLT = twoPathsLeadTo.copy()
                for q in oldTPLT:
                        for u in range(0, self.q):
                                for v in range(0, self.q):
                                        if not (u,v) in
    twoPathsLeadTo:
                                                for b in range(0,
    self.b):
                                                        if (q[0],u,b)
     in t and (q[1],v,b) in t:

    twoPathsLeadTo.add((u,v))
                if oldTPLT == twoPathsLeadTo: break # nothing new was
     added
         return not((qFinal, qFinal) in twoPathsLeadTo)
```

Figure 4.1: Python 3.7 code to test for ambiguity.

`self.q` is the number of states, and `self.b` is the alphabet size. For M to be ambiguous there would have to be [65] a path from the start state to a pair of accept states (in this case $\{q_0, q_0\} = \{q_0\}$) that passes through a pair $\{q_i, q_j\}$ with $i \neq j$.

But $A^-(x) = 5$. See https://math.hawaii.edu/wordpress/bjoern/complexity-of-00011101/. However, a shorter example is the length-6 word 001110 from Theorem 2.19 and Theorem 3.9. It has $A^-(x) = 4$ and $A^{\text{unamb}}(x) = 3$ by being a power-bordered word. □

```
import numpy
import time
import sys
import math

class SimpleStrings:
        def __init__(self,
                    q,               # number of states
                    n,               # length of word
                    b=2,    # alphabet size
                    m=0,    # look for automata that accept at most b^m
    words
                    isItTimeToReturn=False, returnWhenFound=False,
                    edgeLimit=1000 # lower this when calculating E_N
        ):
                    self.q, self.n, self.b, self.m = int(q), int(n), int(
    b), int(m)
                    self.isItTimeToReturn = isItTimeToReturn
                    self.returnWhenFound = returnWhenFound
                    self.edges = [0]*(self.n + 1) # since # edges visited
     <= length of the input string
                    self.edges = numpy.array(self.edges)
                    self.bTom = self.b**m    # better not recalculate this
     over and over...
                    self.lastTime = time.time()
                    self.edgeLimit = edgeLimit
```

Figure 4.2: A class to include Python snippets in.

Remark 4.31. The class framework for the Python code we have shown above is given in Figure 4.2. We then invoke it from a separate file using code like the following:

```
xseq = (0,0,1,0,0,0,0,1)
n=len(xseq)
print("Length: ",n," Word: ",xseq)
b=max(xseq)+1
s = SimpleStrings(q = len(xseq)/2+1, n = len(xseq), b = b, m = 0)
s.complexity(xseq,checkUnambiguity=True,checkDeterminism=False)
```

Theorem 4.32. *There is a word x for which* $A^{\text{unamb}}(x) + 2 = A^{-}(x)$.

Proof. Consider the length-31 word[2] $x =$ `0001010110100001100100111110111`, which is accepted by the NFA in (4.4). We claim that this NFA witnesses that $A^{\text{unamb}}(x) \leq 16$. On the other hand, by a brute force computation we found in [39] that $A^{-}(x) = 18$.

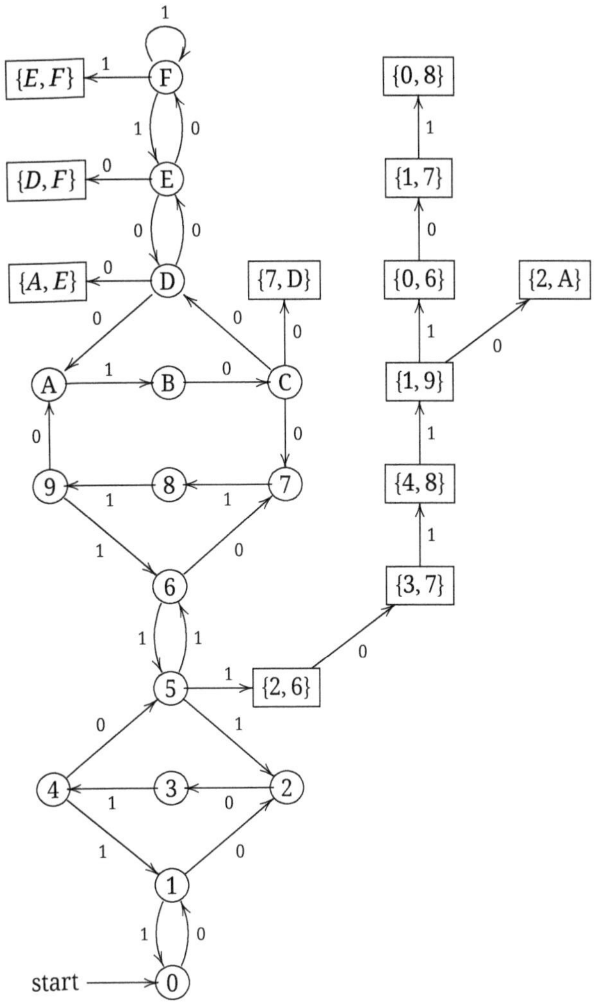

(4.4)

The sets $\{q_i, q_j\}$ of cardinality greater than 1 are indicated with boxes. An arrow leaves a box if two distinct elements of the box can follow that label to the indicated state or set of states. In this example, however, these all "die out." For example, from {2, A} we

2 This is a linear feedback shift register (LFSR) m-sequence [39].

cannot go anywhere since from 2 we can only follow the label 0 and from A we can only follow the label 1. □

Theorem 4.33. *We have* $A^{\text{unamb}} \neq A_N$.

Proof. Let $x = 00100001$. It has the (highly) nondeterministic witnessing state sequence 012333120,[3] representing an NFA with four states and six edges:

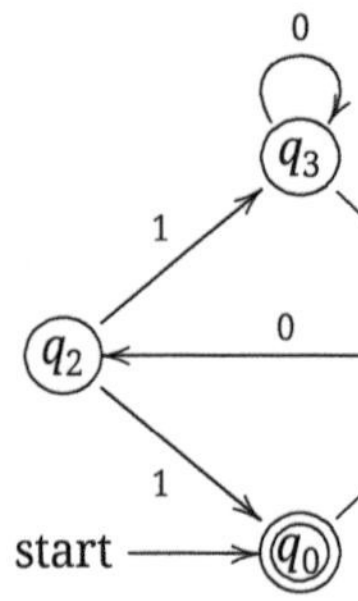

By brute force computation, the word x has no unambiguous witness with four states. □

4.4 Bi-determinism

We can define the bi-deterministic automatic complexity $A^{\pm}$ or groupoid automatic complexity. A more verbose but perhaps more precise taxonomy would go as follows:

$$\begin{aligned} A^{\text{perm}} &= A^{\mathbf{Grp}} \quad \text{(groups)} \\ A^{\pm} &= A^{\mathbf{Grpd}} \quad \text{(groupoids)} \\ A &= A^{\mathbf{Mon}} \quad \text{(monoids)} \\ A^{-} &= A^{\mathbf{PMon}} \quad \text{(partial monoids)} \\ A_{Ne} &= A^{\mathbf{Rel}} \quad \text{(relations).} \end{aligned}$$

In all cases, in defining the complexity of w we are asking for a representation, or action, of a structure generated by δ_a, $a \in \Sigma$, on a finite set, such that $\delta^*(q_0, v) \cap F \neq \emptyset$ for $v \in \Sigma^{|w|}$ iff $v = w$.

An NFA that happens to be deterministic and whose reverse is also deterministic is *bi-deterministic*.

3 Interestingly, the same state sequence 012333120 is thus used to separate A^{unamb} from both A^{-} and A_N, but for different words.

Definition 4.34. Let x be a word. The *bi-deterministic automatic complexity* $A^{\pm}(x)$ is the minimum number of states of a bi-deterministic NFA which uniquely accepts x. The edge-counting version is $E^{\pm}(x)$, the minimum number of edges in a bi-deterministic NFA that uniquely accepts x.

Theorem 4.35. *For any word x, $A^{-}(x^R) \leq A^{\pm}(x)$ and $A^{-}(x) \leq A^{\pm}(x)$.*

Proof. If M is a bi-deterministic NFA and M uniquely accepts x, then M^R uniquely accepts x^R. □

Theorem 4.36. *We have $A^{-} < A^{\pm}$.*

Proof. By Theorem 1.37 and Theorem 4.35,

$$A^{-}(011100) = 3 < 4 \leq A^{-}(001110) = A^{-}(011100^R) \leq A^{\pm}(011100).$$ □

So we have $A_N < A^{-} < A^{\pm}$. We call it $A^{\pm}$ because it may be incomparable to A. When considering $A^{\pm}$ it is natural to draw automata with more symmetry between initial and final states: for example, consider the word $0^2 1^4 0^2$ with the witnessing state sequence 012323201:

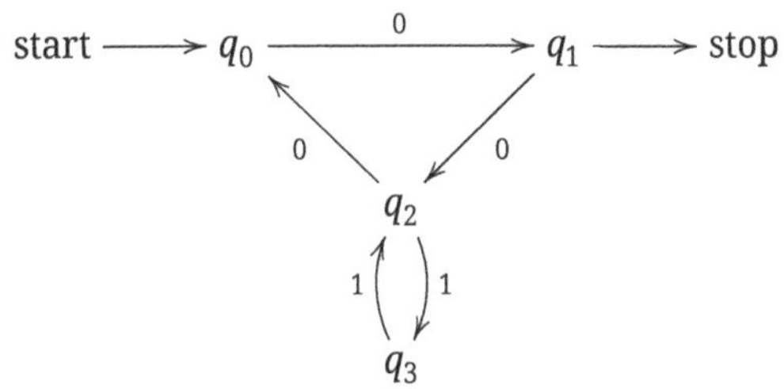

(4.5)

This way of drawing it, it is easy to see that the partial DFA is still deterministic when reversed; in fact, it is isomorphic to its reverse.

Theorem 4.37. *We have $A^{\pm} < A^{perm}$.*

Proof. To see that $A^{\pm} \leq A^{\mathrm{perm}}$, note that if the transition functions restricted to a particular symbol are injective, then on a finite set they are also onto, so that the reverse of the DFA is a DFA.

To see that $A^{\pm} \neq A^{\mathrm{perm}}$, consider $x = \mathtt{01}$; any of the $A_N(x) \leq 2$ witnesses also witness that $A^{\pm}(x) \leq 2$, so we have $A^{\pm}(x) \leq 2 < 3 = |x| + 1 = A^{\mathrm{perm}}(x)$ by Theorem 4.42. □

4.5 Permutation automatic complexity

Question 1.2 suggests defining $A^{\mathrm{tra}}(x)$, the transitive (or irreducible) automatic complexity of x, like $A(x)$ but with the added requirement that there be no dead states. Since two states can form a dead "zone" while being reachable from each other, the natural definition is that each state be reachable from every other state, a kind of "irreducibility" (strong connectivity of the underlying directed graph).

Definition 4.38. The *transitive complexity* A^{tra} and the *doubly transitive complexity* A^{tra2} are defined by modifying the total deterministic automatic complexity A to add the restriction that the monoid generated by $\delta_a, a \in \Sigma$, be transitive and doubly transitive, respectively.

Thus, $A^{\mathrm{tra}}(x)$ is the minimum number of states of a DFA M uniquely accepting x such that for all states q, r of M, q is reachable from r.

Definition 4.39. Let $x \in \Sigma^*$. The *permutation complexity* $A^{\mathrm{perm}}(x, \Sigma)$ is the minimum number of states of a DFA $M = (Q, \Sigma, \delta, q_0, F)$ uniquely accepting x such that for each $a \in \Sigma$, $\delta_a := \delta(\cdot, a)$ is a permutation of Q.

Theorem 4.40. *For all words x, $A^{\mathrm{tra}}(x) \le A^{\mathrm{perm}}(x)$.*

Proof. Let M be a witness for A^{perm} for x. Clearly, each state of M is reachable from the start state q_0. To see that all states are reachable from each other it therefore suffices to show that q_0 is reachable from each state q. For this it suffices to show that if there is an edge from q to r, then q is reachable from r. Let $r = \delta_a(q)$ for some $a \in \Sigma$. Since δ_a is a permutation of the finite set Q, it has some order $k \in \mathbb{N}$, $k \ge 1$. Then

$$q = \delta_a^k(q) = \delta_a^{k-1}(\delta_a(q)) = \delta_a^{k-1}(r). \qquad \square$$

Permutation automatic complexity was first introduced in [38], where Theorem 4.41 was implicitly asserted, but not proved.

Theorem 4.41. *We have $A^{\mathrm{perm}}(x, \Sigma) \le |x| + 1$.*

Proof. A binary word has a decomposition as $0^{n_0}1^{n_1}0^{n_2}\dots$. We include a cycle of length n_0 for 0^{n_0} followed by, and attached to, a cycle of length n_1 for 1^{n_1}, and so on. Symbols not occurring have a self-loop. Stating this precisely is a worthwhile challenge. For the empty word, the unique DFA over Σ on one state is used. For a word $x = x_0$ of length 1, we use the following DFA:

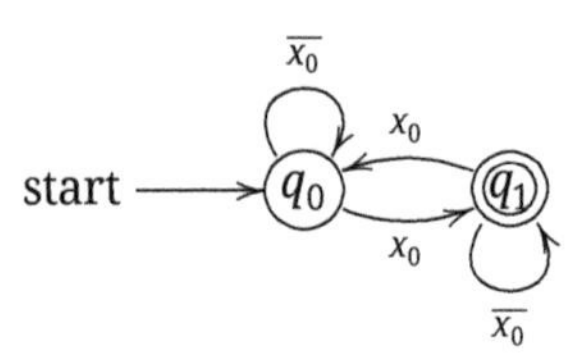

If $x = 0^n$ we define M in two parts:

- The inductive part: $\delta(q_i, 0) = q_{i+1}$ for $0 \le i < n$, $\delta(q_n, 0) = q_0$, and $\delta(q_i, 1) = q_i$ for each $0 \le i < n$.
- The completion part: $\delta(q_n, 1) = q_n$.

If $x = 0^m 1^n$, we let $Q = \{q_0, \ldots, q_{m+n}\}$, with δ extending the inductive part above. In addition, $\delta(q_n, 1) = q_{n+1}$ (inconsistent with the completion part above) and indeed $\delta(q_i, 1) = q_{i+1}$ for each $n \le i < n+m$, $\delta(q_{m+n}, 1) = q_n$, and $\delta(q_i, 0) = q_i$ for each $n+1 \le i \le m$.

For example, for the words `0` and `011`:

`0` :

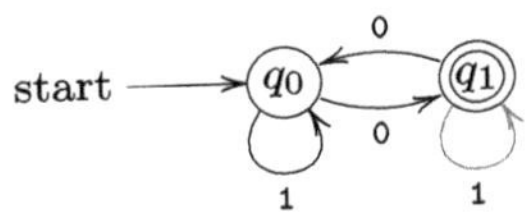

`011` :

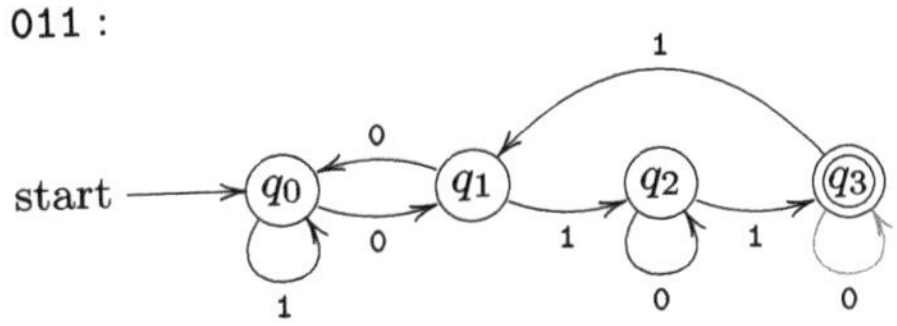

We can express M more directly in permutation group cycle notation. Let $(a_1 \ldots a_k)$ be the cycle π with $\pi(a_i) = a_{i+1}$, $1 \le i < k$, and $\pi(a_k) = a_1$. Write multiplication of such cycles by juxtaposition. Then

$$\delta_0 = (q_0 q_1 \ldots q_{n_0})(q_{n_1} q_{n_1+1} \ldots q_{n_2}) \cdots = \prod_{i=0}^{k} (q_{n_{2i-1}} q_{n_{2i-1}+1} \ldots q_{n_{2i}}),$$

$$\delta_1 = (q_{n_0} q_{n_0+1} \ldots q_{n_1})(q_{n_2} q_{n_2+1} \ldots q_{n_3}) \cdots = \prod_{i=0}^{k} (q_{n_{2i}} q_{n_{2i}+1} \ldots q_{n_{2i+1}}).$$

Here $n_{-1} = 0$. □

Theorem 4.42 (Starting from: Anthony Quas [62]). *If $|\Sigma| \ge 2$, then $A_\Sigma^{\mathrm{perm}}(x) \ge |x| + 1$ for all $x \in \Sigma^n$, $n \ge 0$. If $1 \le |\mathcal{F}| \le |\Sigma| - 1$, $\mathcal{F} \subseteq \Sigma^n$, then $A^{\mathrm{perm}}(\mathcal{F}) \ge n + 1$.*

Proof. We must show the following:

Suppose $\delta_b : Q \to Q$, $b \in \Sigma$, are invertible functions, where $Q = \{0, \ldots, q-1\}$ is a set of q elements. For a word $w = w_1 \ldots w_n \in \Sigma^n$ we define $\delta_w = \delta_{w_n} \circ \delta_{w_{n-1}} \circ \cdots \circ \delta_{w_1}$. Suppose c, d and n are such that $|\{w \in \Sigma^n \mid \delta_w(c) = d\}| = f$, $1 \le f \le |\Sigma| - 1$. Then $q \ge n + 1$.

To prove it, fix $c \in Q$ and $m \ge 0$. Let $\mathcal{R}_m(c) = \{\delta_w(c) : |w| = m\}$ and $r_m(c) = |\mathcal{R}_m(c)|$. In particular, $\mathcal{R}_0(c) = \{c\}$.

Claim 4.43. *Let $m \geq 0$. If $r_m(c) = r_{m+1}(c)$, then for all $M > m$ and for all $d \in \mathcal{R}_M(c)$, there are at least $|\Sigma|$ many words w in Σ^M with $\delta_w(c) = d$.*

Proof. Let $S = \mathcal{R}_m(c)$. Then $\mathcal{R}_{m+1}(c) = \bigcup_{b\in\Sigma} \delta_b(S)$. By invertibility of the δ_b, each $\delta_b(S)$ has cardinality $r_m(c)$, and their union also has cardinality $r_{m+1}(c)$. In our case, $r_{m+1}(c) = r_m(c)$, so it follows that $\delta_b(S) = \delta_{b'}(S)$ for all $b, b' \in \Sigma$.

Now suppose $w_1, \ldots, w_s$, $s < |\Sigma|$, are the words of length M with $\delta_{w_i}(c) = d$, $w_i \in \Sigma^M$. Decompose w_i as $w_i = u_i v_i$, where $|u_i| = m + 1$ and $|v_i| = M - (m + 1) \geq 0$. Let b_i be the last symbol in u_i and let $b' \in \Sigma \setminus \{w_1, \ldots, w_s\}$, $b' \in \Sigma$.

Since $\delta_{b_i}(S) = \delta_{b'}(S)$ for each i, for each i_0, since $\delta_{u_{i_0}}(c) \in \delta_{b_{i_0}}(S) = \delta_{b'}(S)$, there is some u' (depending on i_0) ending in b' with $\delta_{u'}(c) = \delta_{u_{i_0}}(c)$. This u' of length $m + 1$ has $u'(m) \neq u_i(m)$ for each i since u' ends in b'. Now $\delta_v \circ \delta_{u'}(c) = \delta_v \circ \delta_{u_i}(c)$, so $w := v \circ u'$ is another word of length M with $\delta_w(c) = d$ and the claim is proved. □

It follows that if c and d are such that there are at most $|\Sigma| - 1$ many w of length n with $\delta_w(c) = d$, then $r_{j+1}(c) > r_j(c)$ for each $j < n$, so that $q \geq r_n(c) \geq n + r_0(c) = n + 1$. □

The proof of Theorem 4.42 somewhat resembles that of the Morse–Hedlund theorem [58]. By Theorem 4.42 a family of two words of length n over $\{0, 1, 2\}$ must have complexity at least $n + 1$. For $|\mathcal{F}| = 1$, Theorem 4.42 is sharp, as shown in Theorem 4.41. For $|\mathcal{F}| = 2$, for $n \leq 1$ it is sharp but for $n = 2$ it is not: it is tedious to write down, but $\mathcal{F} = \{00, 11\}$ with $\Sigma = \{0, 1, 2\}$ requires at least four states. Actually four states are enough: let δ_2 be the identity and let δ_0 and δ_1 be as indicated in (4.6):

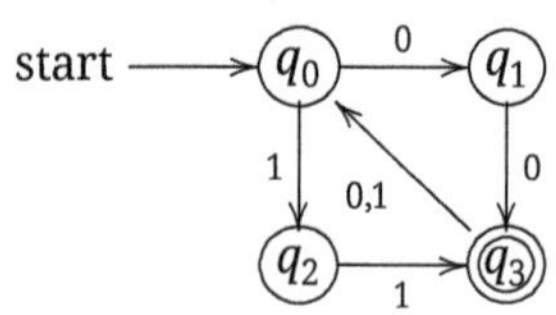

(4.6)

Extending this it seems that $2n$ states are enough when $|\mathcal{F}| = 2$ and in general $|\mathcal{F}| \cdot (n - 1) + 2$ states are enough, which would be a nice generalization of Theorem 4.41, but this requires checking. For $|\mathcal{F}| = 2$ the typical situation is as follows:

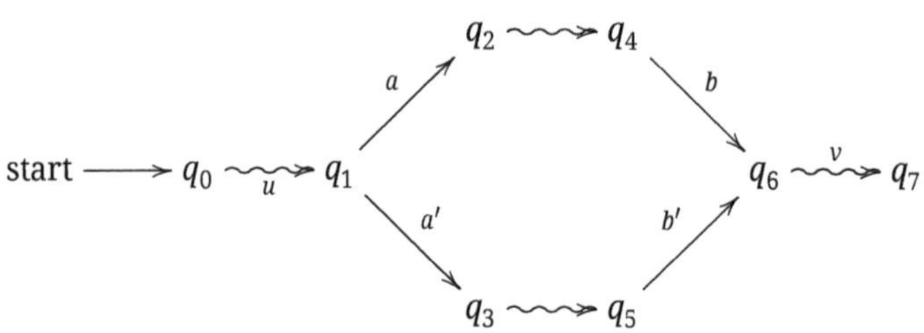

The definition of the DFA M has four cases, depending on whether the last symbol of u is a or $a' \neq a$ and whether the first symbol of v is b or $b' \neq b$. For instance, say $u = u'\vec{a}$ ends with a segment $\vec{a}$ of a's and $v = \vec{b'}v'$ starts with a segment $\vec{b}$ of b''s. Then we would do this (where we have also gathered the [possibly] multiple a's and a''s following q_1 [and similarly for b and b'] into vectors):

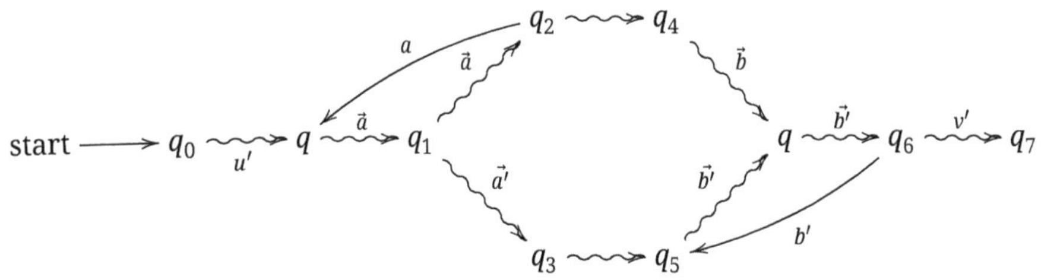

For $\mathcal{F} = 3$ and beyond, this picture does not generalize. In fact, we run into trouble if we insist on a single accept state and a single start state: under these conditions, no group structure exists for $\mathcal{F} = \{00, 01, 11\}$. On the other hand, if we allow multiple accept states we can use the complementary DFA to the one for $\mathcal{F} = \{10\}$. The following general upper bound holds and specializes to Theorem 4.41 when $k = 0$.

Definition 4.44. Let Σ be a finite alphabet. Let $n \in \mathbb{N}$ and let $\mathcal{F} \subseteq \Sigma^n$. $A^{\text{perm}}(\mathcal{F}, \Sigma)$ is the minimum number of states of a DFA over Σ for which each $\delta_a = \delta(\cdot, a)$, $a \in \Sigma$, is injective, with a single start state and multiple final states allowed, such that $L(M) \cap \Sigma^n = \mathcal{F}$.

Theorem 4.45. *Let $n \in \mathbb{N}$ and let $\emptyset \neq \mathcal{F} \subseteq \{0,1\}^n$. Let k be such that $|\Sigma|^{k-1} < \mathcal{F} \leq |\Sigma|^k$. Then*

$$A^{\text{perm}}(\mathcal{F}, \Sigma) \leq \frac{|\Sigma|^k - 1}{|\Sigma| - 1} + |\mathcal{F}| \cdot (n + 1 - k).$$

Proof. We use a tree of states. In the worst case, the elements of $\mathcal{F}$ split off early. In that case there are

$$|\mathcal{F}| + \sum_{i=0}^{k-1} |\Sigma|^i = \frac{|\Sigma|^{(k-1)+1} - 1}{|\Sigma| - 1} + |\mathcal{F}|$$

states needed for the lengths up to the last splitting off point, and $|\mathcal{F}| \cdot (n - k)$ further states after that. For $\Sigma = \{0, 1\}$ and $|\mathcal{F}| = 6$ we could have:

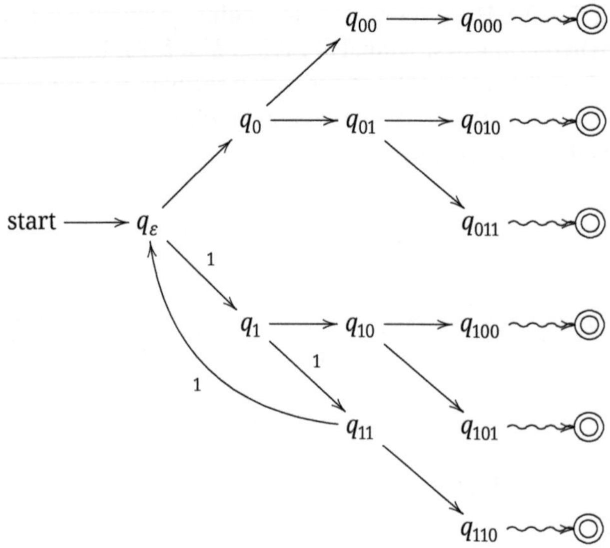

For each $a \in \Sigma$, δ_a is a product of the disjoint cycles

$$(q_w, q_{wa}, q_{waa}, \dots, q_{wa^t}),$$

where wa^t is a prefix of a member of $\mathcal{F}$ and wa^{t+1} is not and where w does not end in a. □

Theorem 4.46. *We have $A^{\text{tra}} \neq A^{\text{perm}}$.*

Proof. By Theorem 4.42 it suffices to find an x with $A^{\text{tra}}(x) \leq |x|$. Here is an example showing we can take any x with $|x| = 3$:

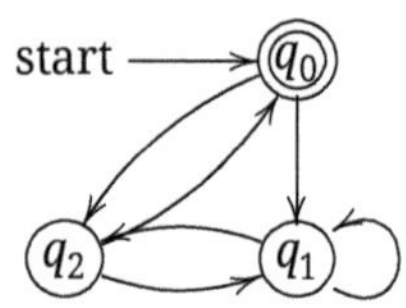

Add labels `0`, `1` arbitrarily to make this a total DFA over {`0`, `1`}. For instance, for a word $x_0x_1x_2$:

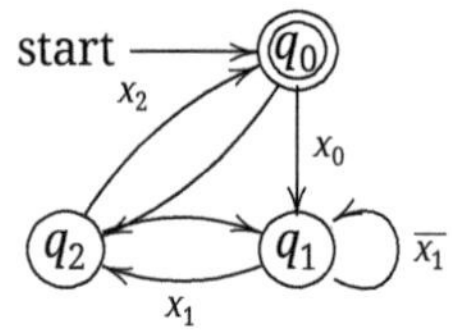

To see that there is only one accepted word of length 3, note that if $\delta(q_0, x_0x_1x_2) = q_0$, then $\delta(q_0, x_0x_1) = q_2$ and hence $\delta(q_0, x_0) \in \{q_0, q_1\}$. But $\delta(q_0, x_0) = q_0$ is impossible, so there is a unique walk. Alternatively, the transition matrix $A = [a_{ij}]$, where a_{ij} is the number of walks from q_{i-1} to q_{j-1}, is

$$A = \begin{pmatrix} 0 & 1 & 1 \\ 0 & 1 & 1 \\ 1 & 1 & 0 \end{pmatrix}, \quad A^3 = \begin{pmatrix} 1 & 4 & 3 \\ 1 & 4 & 3 \\ 2 & 4 & 2 \end{pmatrix},$$

in particular $a_{11} = 1$. □

We can generalize the proof of Theorem 4.46 as follows.

Theorem 4.47. *Let Σ be an alphabet and let $x \in \Sigma^*$ with $|x| \geq 3$. Then $A^{\text{tra}}(x) \leq |x|$.*

Proof. Let $x = x_0 \ldots x_{n-1}$ with each $x_i \in \Sigma$. Let $M = (Q, \Sigma, \delta, q_0, \{q_0\})$, where $Q = \{q_0, \ldots, q_{n-1}\}$, $|Q| = n$, and δ is defined as follows. For each $2 \leq k \leq n-1$, we let $\delta(q_k, x_k) = q_{k+1}$ and $\delta(q_k, a) = q_{k-1}$ for each $a \in \Sigma \setminus \{x_k\}$. The only other states are q_0 and q_1, which have $\delta(q_0, x_0) = q_1$, $\delta(q_1, x_1) = q_2$, $\delta(q_0, a) = q_2$ for any $a \neq x_0$ and $\delta(q_1, a) = q_1$ for any $a \neq x_1$.

For $n = 7$ and $\Sigma = \{0, 1\}$, M is shown in (4.7), where missing labels are to be filled in so that a complete DFA is formed:

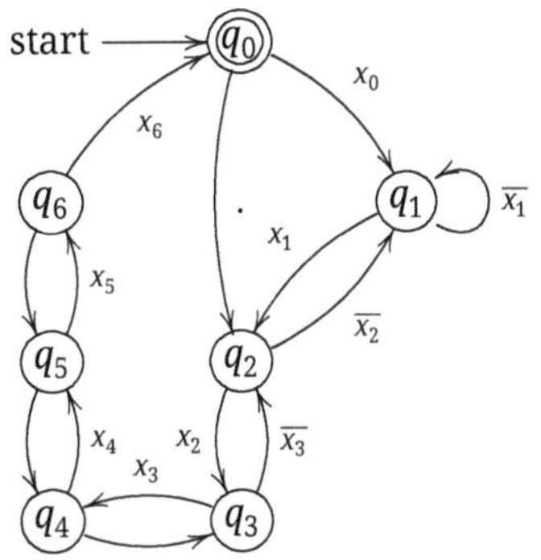

(4.7)

Clearly, M accepts x. For a word y of length n, if $\delta''y \upharpoonright n = q_0$ (we are at q_0 at time n), then $\delta''y \upharpoonright (n-1) = q_{n-1}$ (we must be at q_{n-1} at time $n-1$). The latest time we could be at q_2 is time 2. Being at q_0 at time 1 is inconsistent with being at q_0 at time 0. Therefore, we must have been at q_1 at time 1, so the whole walk is determined. □

Remark 4.48. We can get a little more information from the proof of Theorem 4.46. The transition matrix is

$$\begin{pmatrix} 0 & 1 & 1 & 0 & 0 & 0 & 0 \\ 0 & 1 & 1 & 0 & 0 & 0 & 0 \\ 0 & 1 & 0 & 1 & 0 & 0 & 0 \\ 0 & 0 & 1 & 0 & 1 & 0 & 0 \\ 0 & 0 & 0 & 1 & 0 & 1 & 0 \\ 0 & 0 & 0 & 0 & 1 & 0 & 1 \\ 1 & 0 & 0 & 0 & 0 & 1 & 0 \end{pmatrix}$$

and the NFA is shown in (4.8). Raising this matrix to the power 8 we see that there is a unique walk of length 8 from q_2 to q_0. Thus, we can achieve $A^{\text{tra}}(x) \le 7$ for a word x of length 8 and a unique walk from q_3 to q_0 of length 9. The word of length 9 must be of the form $x_0x_1x_2\overline{x_2x_1x_0}x_6x_7x_8$, that is, a word of length 3, its complemented reverse, and another word of length 3:

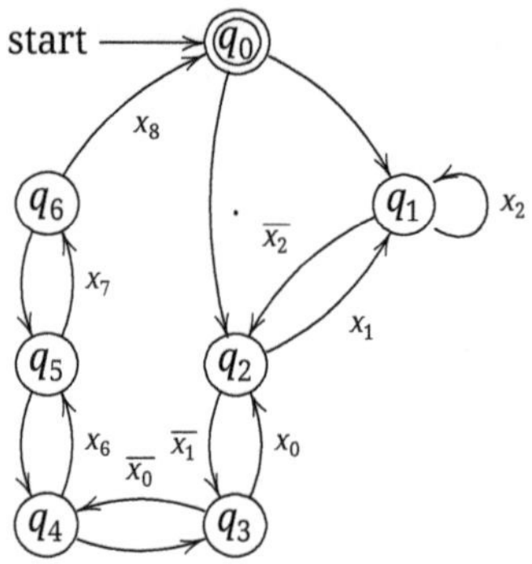

(4.8)

In Theorem 4.49 we extend Theorem 4.46 to obtain a doubly transitive witness.

Theorem 4.49. *We have* $A^{\text{tra2}} < A^{\text{perm}}$.

Proof. We first prove $A^{\text{tra2}} \le A^{\text{perm}}$. Our witnesses for $A^{\text{perm}}(x) \le |x| + 1$ are doubly transitive since the generated group is S_n. Therefore, $A^{\text{tra2}}(x) \le |x| + 1 = A^{\text{perm}}(x)$.

We now prove $A^{\text{tra2}}(x) < |x| + 1$ for some x. Consider the three-state witness given in Theorem 4.46 with the accepted word $x_0x_1x_2 = 000$. This is doubly transitive in the weak (group) sense that for any $q_i \ne q_j$ and $q_k \ne q_l$, there is a function f in the generated monoid with $f(q_i) = q_k$ and $f(q_j) = q_l$. It is in fact also doubly transitive in the strong sense where we do not require $q_k \ne q_l$, since the monoid contains all the constant functions.

1-Transitivity is shown by the following computation:

	0	**1**	**2**
id	0	1	2
δ_0	1	2	0
δ_1	2	1	1
δ_0^2	2	0	1

To get 2-transitivity we continue as follows:

	0	1	2
$\delta_0\delta_1$	0	2	2
$\delta_1\delta_0$	1	1	2
δ_1^2	1	1	1
$\delta_0^2\delta_1$	1	0	0
$\delta_0\delta_1\delta_0$	2	2	0
$\delta_0\delta_1^2$	2	2	2
$\delta_1\delta_0^2$	1	2	1
$\delta_0^2\delta_1\delta_0$	0	0	1
$\delta_0^2\delta_1^2$	0	0	0
$\delta_0\delta_1\delta_0^2$	2	0	2
$\delta_1\delta_0^2\delta_1$	1	2	2
$\delta_0^2\delta_1\delta_0^2$	0	1	0

□

4.6 Counter automata

A pushdown automaton is formally defined as a 7-tuple $M = (Q, \Sigma, \Gamma, \delta, q_0, Z, F)$, where:
- Q is a finite set of states;
- Σ is a finite set which is called the input alphabet;
- Γ is a finite set which is called the stack alphabet;
- δ is a finite subset of $Q \times (\Sigma \cup \{\varepsilon\}) \times \Gamma \times Q \times \Gamma^*$, the transition relation;
- $q_0 \in Q$ is the start state;
- $Z \in \Gamma$ is the initial stack symbol; and
- $F \subseteq Q$ is the set of accepting states.

An element $(p, a, A, q, \alpha) \in \delta$ is a transition of M. It has the intended meaning that M, in state $p \in Q$, on the input $a \in \Sigma \cup \{\varepsilon\}$ and with $A \in \Gamma$ as topmost stack symbol, may read a, change the state to q, and pop A, replacing it by pushing $\alpha \in \Gamma^*$. The $(\Sigma \cup \{\varepsilon\})$ component of the transition relation is used to formalize that the PDA can either read a letter from the input or proceed leaving the input untouched. A counter automaton, or counter machine, is a pushdown automaton with only two symbols, A and the initial symbol, in Γ, the finite set of stack symbols.

Definition 4.50. The 1-counter automatic complexity A_C (Figure 4.3) is defined as follows. For 1-counter DFAs and NFAs, we may increment or decrement a counter while reading each symbol. If the counter, which starts at 0, ever becomes negative, then the computation hangs and the word is not accepted.

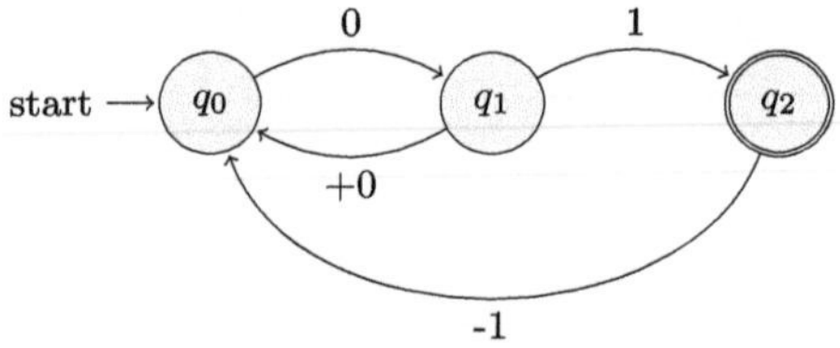

Figure 4.3: An automaton witnessing $A'_C(0001101) \leq 3$.

If we additionally require that the counter be returned to 0 at the end of the computation, then we get a modified 1-counter automatic complexity A'_C.

In both cases we do not allow ε-transitions. If ε-transitions are desired (including modifying the stack without reading the input) we may speak of A^{ε}_C and $A^{\varepsilon'}_C$.

Note that $A_C(x) \leq A'_C(x)$ for each x.

Remark 4.51. Shallit and Breitbart [68] explored a notion of descriptional complexity, similar to automatic complexity, for languages that is uninteresting for the case of a single string. A few words at this point about models of computation. The powerful Turing machines are the basis for the noncomputable Kolmogorov complexity. Between Turing machines and the simpler finite automata, one can also find fertile ground. Consider a context-free grammar in Chomsky normal form: all productions are of the form $A \to BC$ or $A \to a$, where A, B, C are variables and a is a terminal. Use the number of variables as the measure of a grammar's size. This gives a well-known measure of complexity associated with "word chains," studied by Diwan [17] and subsequently in [2, 3, 6, 9, 64].

4.7 Open problems

At length 2 we have $A^{\mathrm{tra}}(01) = 2$ but $A^{\mathrm{tra}}(00) = 3$. Since A^{perm} is very efficiently computable by Theorem 4.42, it is interesting to ask the following question.

Question 4.1. Is A^{tra} efficiently computable (say, in polynomial time)? Are any of the functions in the master diagram (4.1) that are not indicated as efficiently computable there, like A itself, or A_N, efficiently computable?

Question 4.2. The monoid for the DFA in Theorem 4.49 has the structure shown in (4.9) with 24 elements. Note that it is not 3-transitive, since it does not contain all permutations; in fact the only permutations it does contain are the powers of δ_0. We fairly clearly see the structure of ideals: the minimal ideal generated to 111 and the further nontrivial ideals generated by 121, 112, and 211, respectively. What does the structure of ideals look like for, say, a Kayleigh graph?

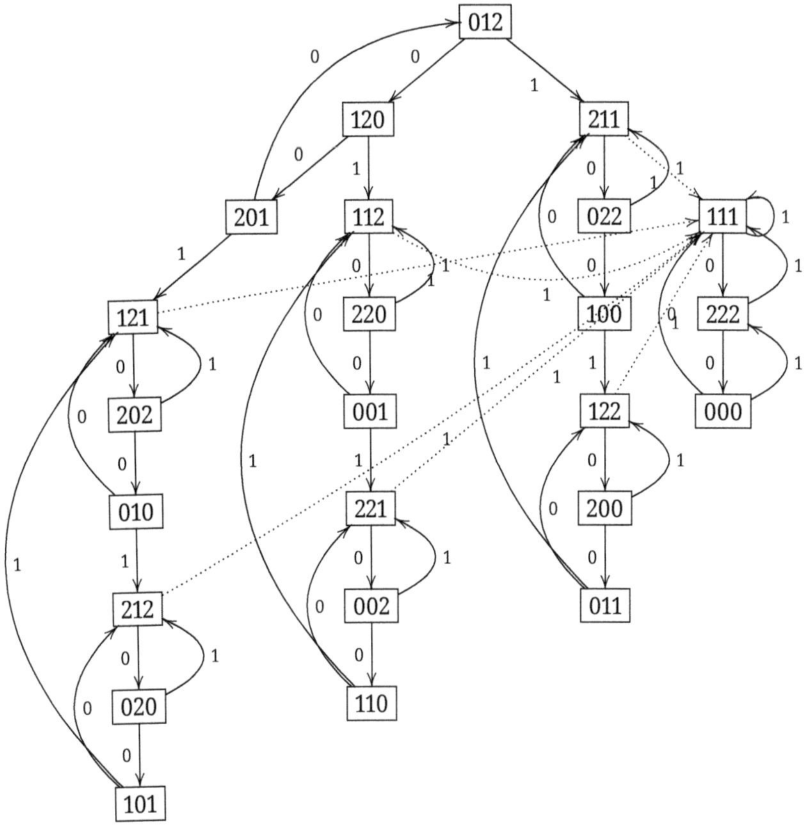

(4.9)

Question 4.3. Is A_N level-convex (Definition 4.10)?

Question 4.4. Are the differences $A^-(y) - A_N(y)$ unbounded as $|y| \to \infty$?

Question 4.5. In Definition 4.50, is $A_C \neq A'_C$? That is, does there exist a word x with $A_C(x) < A'_C(x)$?

Question 4.6. Does either A_C or A'_C satisfy the subword inequality (1.7)?

Question 4.6 may be easier to answer in the negative for A'_C, since then 0^n1^n should be simpler than 0^k0^l for some $k, l < n$, than for A_C, A^{ε}_C, or $A^{\varepsilon'}_C$.

4.8 Exercises

Exercise 4.1. Verify that the Collatz sequence [75] begins as follows: 0, 1, 7, 2, 5, 8, 16, 3, 19, 6, 14, 9, 9.

A complete solution to Exercise 4.1 is available at the following URL.

https://github.com/bjoernkjoshanssen/ac-exercises/

Exercise 4.2. Let G be a simple graph and let S be a set of vertices of G. If there exists a walk from an element of S to some vertex not in S, then some vertex in S is adjacent to some vertex not in S.

A complete solution to Exercise 4.2 is available at the following URL.

https://github.com/bjoernkjoshanssen/ac-exercises/

Exercise 4.3. Let $y = \texttt{1100012101011011110} \in \{\texttt{0}, \texttt{1}, \texttt{2}\}^{19}$ and $x = \texttt{110001101011011110} \in \{\texttt{0}, \texttt{1}\}^{18}$. Let π be the morphism

$$\pi(0) = 0, \quad \pi(1) = 1, \quad \pi(2) = \varepsilon.$$

Check that $x = \pi(y)$.

A complete solution to Exercise 4.3 is available at the following URL.

https://github.com/bjoernkjoshanssen/ac-exercises/

5 The incompressibility theorem

Shallit and Wang showed that the automatic complexity $A(x)$ satisfies $A(x) \geq n/13$ for almost all $x \in \{0,1\}^n$. They also stated that Holger Petersen had informed them that the constant 13 can be reduced to 7. Here we show that it can be reduced to $2+\epsilon$ for any $\epsilon > 0$. The result also applies to nondeterministic automatic complexity $A_N(x)$. In that setting the result is tight inasmuch as $A_N(x) \leq n/2+1$ for all x. Thus we show (Theorem 5.14) that automatic complexity has a similar incompressibility phenomenon to that of Kolmogorov complexity for Turing machines, first studied in [47, 48, 77, 78].

5.1 Incompressibility

Let C denote Kolmogorov complexity, so that $C(\sigma)$ is the length of the shortest program, for a fixed universal Turing machine, that outputs σ on empty input. Let $\omega = \{0,1,2,\ldots\}$ be the set of nonnegative integers and let $\omega^{<\omega} = \omega^*$ be the set of finite words over ω.

As Solomonoff and Kolmogorov observed, for each n there is a word $\sigma \in \{0,1\}^n$ with $C(\sigma) \geq n$. Indeed, each word with $C(\sigma) < n$ uses up a description of length $< n$, and there are at most $\sum_{k=0}^{n-1} 2^k = 2^n - 1 < 2^n = |\{0,1\}^n|$ of those.

Similarly, we have the following lemma.

Lemma 5.1 (Solomonoff, Kolmogorov). *For each nonnegative integer n, there are at least $2^n - (2^{n-k}-1)$ binary words σ of length n such that $C(\sigma) \geq n-k$.*

Proof. For each word with $C(\sigma) < n-k$ we use up at least one of the at most $2^{n-k}-1$ many possible descriptions of length less than $n-k$, leaving at least

$$|\{0,1\}^n| - (2^{n-k}-1)$$

words σ that must have $C(\sigma) \geq n-k$. □

Shallit and Wang connected their automatic complexity $A(x)$ with Kolmogorov complexity in the following Theorem 5.2.

Theorem 5.2 (Shallit and Wang [69, proof of Theorem 8]). *For all binary words x,*

$$C(x) \leq 12A(x) + 3\log_2|x| + O(1).$$

They mention ([69, proof of Theorem 8]), without singling it out as a lemma, the following result (Lemma 5.4). Since they used, but did not give a definition of, the notion of *almost all*, we give a definition here. The notion is also known by the phrase *natural density* 1.

https://doi.org/10.1515/9783110774870-005

Definition 5.3. A set of strings $S \subseteq \{0,1\}^*$ contains almost all $x \in \{0,1\}^n$ if

$$\lim_{n\to\infty} \frac{|S \cap \{0,1\}^n|}{2^n} = 1.$$

Lemma 5.4. *We have* $C(x) \geq |x| - \log_2|x|$ *for almost all* x.

Proof. Let $S = \{x \in \{0,1\}^* : C(x) \geq |x| - \log_2|x|\}$. By Lemma 5.1,

$$\lim_{n\to\infty} \frac{|S \cap \{0,1\}^n|}{2^n} \geq \lim_{n\to\infty} \frac{2^n - (2^{n-\log_2 n} - 1)}{2^n} = \lim_{n\to\infty} 1 - \left(\frac{1}{n} - \frac{1}{2^n}\right) = 1.$$

□

Shallit and Wang then deduced the following.

Theorem 5.5 ([69, Theorem 8]). *For almost all* $x \in \{0,1\}^n$ *we have* $A(x) \geq n/13$.

Proof. By Lemma 5.4 and Theorem 5.2 there is a constant C such that for almost all x,

$$|x| - \log_2|x| \leq C(x) \leq 12A(x) + 3\log_2|x| + C.$$

Let $C' = C/12$. By taking n large enough,

$$\frac{n}{13} \leq \frac{n}{12} - \frac{1}{3}\log_2 n - C' \leq A(x).$$

□

Our main result (Theorem 5.14) implies that for all $\epsilon > 0$, $A(x) \geq n/(2+\epsilon)$ for almost all words $x \in \{0,1\}^n$. Analogously, one way of expressing the Solomonoff–Kolmogorov result is as follows.

Proposition 5.6. *For each* $\epsilon > 0$, *the following statement holds:* $C(x) \geq |x|(1-\epsilon)$ *for almost all* $x \in \{0,1\}^n$.

The core idea for Theorem 5.14 is as follows. Consider an automaton processing a word x of length n over $n+1$ points in time. We show that there exist powers $x_i^{\alpha_i}$ within x with $\alpha_i \geq 2$ and all distinct base lengths $|x_i|$, which in total occupy $\sum 1 + \alpha_i|x_i|$ time, such that all other states are visited at most twice. Since most words do not contain any long powers, this forces the number of states to be large.

5.2 Deepening the power–complexity connection

We have seen connections between occurrences of powers in words and their automatic complexity in Theorem 2.6. It turns out that we can get more mileage out of this vehicle.

The reader may note that in the context of automatic complexity, Definition 1.15 can without loss of generality be simplified as follows:

1. We may assume that the set of final states is a singleton.
2. We may assume that whenever $q, r \in Q$ and $b_1, b_2 \in \Sigma$, if $r \in \delta(q, b_1) \cap \delta(q, b_2)$, then $b_1 = b_2$. Indeed, having multiple edges from q to r in an automaton witnessing the automatic complexity of a word would violate uniqueness.
3. Each automaton M may be assumed to be *generated by a witnessing walk*. That is, only edges used by a walk taken when processing x along the unique accepting walk need to be included in M.

Let us call an NFA M *witness-generated* if there is some $x \in \Sigma^*$ such that x is the only word of length $|x|$ that is accepted by M, M accepts x along only one walk, and every state and transition of M is visited during this one walk. In this case we also say that M is witness-generated by x. When studying nondeterministic automatic complexity we may, without loss of generality, restrict our attention to witness-generated NFAs.

In Theorem 5.7 we refer to Definition 1.4 and Definition 1.16.

Theorem 5.7. *Let n be a positive integer. Let $D = (V, E)$ be a digraph and let $s, t \in V$. Suppose that there is a unique walk Δ on D from s to t of length n and that for each $e \in E$ there is a t with $(\Delta(t), \Delta(t+1)) = e$. Then there is a set of disjoint cycles $\mathcal{C}$ such that*

$$v \in V \setminus \bigcup_{C \in \mathcal{C}} \text{range}(C) \quad \Longrightarrow \quad \big|\{t : \Delta(t) = v\}\big| \le 2$$

and such that for each $C \in \mathcal{C}$ there exist $\mu_C \ge 2|C|$ and t_C such that

$$\begin{aligned} \{t : \Delta(t) \in \text{range}(C)\} &= [t_C, t_C + \mu_C] \quad \textit{and} \\ \Delta(t_C + k) &= C(k \bmod |C|) \quad \textit{for all } 0 \le k \le \mu_C. \end{aligned} \tag{5.1}$$

Proof. Suppose $v \in V$ with $\{t : \Delta(t_j) = v\} = \{t_1 < t_2 < \cdots < t_k\}$ and $k \ge 3$. Let us write $\Delta_{[a,b]}$ for the sequence $(\Delta(a), \dots, \Delta(b))$ for any a, b.

Claim. *The vertex sequence $S = \Delta_{[t_j, t_{j+1}]}$ does not depend on j.*

Proof of claim. For $k = 3$, $v \in V$ with $\Delta(t_1) = \Delta(t_2) = \Delta(t_3)$, for some $t_1 < t_2 < t_3$. Then the same vertex sequence must have appeared in $[t_1, t_2]$ and $[t_2, t_3]$:

$$\Delta_{[t_1,t_3]} = \Delta_{[t_2,t_3]} +\!+ \Delta_{[t_1+1,t_2]},$$

or else uniqueness of walk would be violated since

$$\Delta_{[0,t_1-1]} +\!+ \Delta_{[t_2,t_3]} +\!+ \Delta_{[t_1+1,t_2]} +\!+ \Delta_{[t_3+1,n]}$$

would be a second walk on D from s to t of length n. For $k > 3$ the only difference in the argument is notational. □

By definition of the t_j's, S is a cycle except for reindexing. Thus, let $C(r) = S(t_1 + r)$ for all r, let $t_C = t_1$, and let $\mu = \mu_C$ be defined by (5.1). We have

$$t_C + \mu_C \geq t_k = t_1 + \sum_{j=1}^{k-1} t_{j+1} - t_j = t_1 + (k-1)|C|$$

and hence $\mu_C \geq (k-1)|C| \geq 2|C|$. □

Definition 5.8. Let M be an NFA. The directed graph $D(M)$ has the set of states Q as its set of vertices and has edges (s,t) whenever $t \in \delta(s,b)$ for some $b \in \Sigma$.

Theorem 5.9. *Let $q \geq 1$, $n \geq 0$, and let x be a word of length n such that $A_N(x) = q$. Then x contains a set of powers $x_i^{\alpha_i}$, $\alpha_i \geq 2$, $1 \leq i \leq m$, satisfying* (5.2) *and* (5.3) *with $\beta_i = \lfloor \alpha_i \rfloor$:*

$$\sum_{i=1}^{m} \beta_i |x_i| = \sum_{i=1}^{m} \gamma_i |x_i|, \quad \gamma_i \in \mathbb{Z}, \gamma_i \geq 0 \quad \Longrightarrow \quad \gamma_i = \beta_i, \quad \textit{for each } i, \tag{5.2}$$

$$n + 1 - m - \sum_{i=1}^{m} (\alpha_i - 2)|x_i| \leq 2q. \tag{5.3}$$

Theorem 5.9 appears as [42, Theorem 15].

Proof. Let M be an NFA witnessing that $A_N(x) \leq q$, with digraph $D = D(M)$. Let $\mathcal{C} = \{C_1, \ldots, C_m\}$ be a set of disjoint cycles in D as guaranteed by Theorem 5.7. Then let x_i be the word read by M while traversing C_i and let $\alpha_i = \mu_{C_i}$ from Theorem 5.7.

Since the C_i are disjoint, there are $\Omega := q - \sum_{i=1}^{m} |x_i|$ vertices not in $\bigcup_i C_i$. Let $P = |\{t : \Delta(t) \in C_i, \text{for some } i\}|$ and let $N = n + 1 - P$. By Theorem 5.7, $N \leq 2\Omega$, so $P = n + 1 - N \geq n + 1 - 2\Omega$. On the other hand, $P = \sum_{i=1}^{m}(1 + \alpha_i |x_i|)$, since a walk of length k is the range of a function with domain of cardinality $k+1$. Substituting back into the inequality $P \geq n + 1 - 2\Omega$ now yields

$$\sum_{i=1}^{m} (1 + \alpha_i |x_i|) \geq n + 1 - 2\left(q - \sum_{i=1}^{m} |x_i| \right)$$

and hence (5.3). □

Theorem 5.10. *Let $q \geq 1$ and let x be a word such that $A_N(x) \leq q$. Then x contains a set of powers $x_i^{\alpha_i}$, $\alpha_i \geq 2$, $1 \leq i \leq m$, such that all the $|x_i|$, $1 \leq i \leq m$, are distinct and nonzero and satisfy* (5.3).

Proof. This follows from Theorem 5.9 once we note that unique solvability of (5.2) implies that all the lengths are distinct.

The unique solution is $\beta_k = \lfloor \alpha_k \rfloor \geq 1$. Suppose $|x_i| = |x_j|$, $i \neq j$. Then another solution is $\gamma_k = \beta_k$ for $k \notin \{i,j\}$, $\gamma_i = \beta_i - 1$, $\gamma_j = \beta_j + 1$. □

For a word $x = x_1 \ldots x_n$ with each $x_i \in \{0, 1\}$ we write $x_{[a,b]} = x_a x_{a+1} \ldots x_b$.

Definition 5.11. Let $x = x_0 \dots x_{n-1}$ with each $x_i \in \{0, 1\}$. Lookback(m, k, t, x) is the statement that $x_{m+u} = x_{m+u-k}$ for each $0 \le u < t$, i. e.,

$$\text{Lookback}(m, k, t, x) \iff x_{[m:m+t-1]} = x_{[m-k:m+t-1-k]}.$$

Example 5.12. For a word $x = x_0 \dots x_4 \in \{0, 1\}^5$, we have Lookback$(2, 1, 1, x)$ iff $x_1 = x_2$ (see Exercise 5.1).

We can read Lookback(m, k, t, x) as "position m starts a continued run with lookback amount k of length t in x."

In Lemma 5.13 we state a form of the pigeonhole principle to be used in Theorem 5.14.

Lemma 5.13. *If m many distinct positive integers $y_1, \dots, y_m$ are all bounded by z, then it follows that $m \le z$.*

Proof. The map $i \mapsto y_i$ is an injection from $\{1, \dots, m\}$ into $\{1, \dots, z\}$. □

Theorem 5.14. *Let $\mathbb{P}_n$ denote the uniform probability measure on words $x \in \Gamma^n$, where Γ is a finite alphabet of cardinality at least 2. For all $\epsilon > 0$,*

$$\lim_{n\to\infty} \mathbb{P}_n\left(\left|\frac{A_N(x)}{n/2} - 1\right| < \epsilon\right) = 1.$$

Proof. Let us write $\log = \log_{|\Gamma|}$ in this proof. Let $d = 3$, although any fixed real number $d > 2$ will do for the proof. For $1 \le m \le n$ and $1 \le k \le m$ let $R_{m,k} = \{x \in \Gamma^n :$ Lookback$(m, k, \lceil d \log n\rceil, x)\}$. By the union bound,[1]

$$\mathbb{P}_n\left(\bigcup_{m=1}^{n}\bigcup_{k=1}^{m} R_{m,k}\right) \le \sum_{m=1}^{n}\sum_{k=1}^{m} |\Gamma|^{-d\log n} = n^{-d}\sum_{m=1}^{n} m = \frac{n(n+1)}{2}\cdot n^{-d} =: \epsilon_{n,d}. \tag{5.4}$$

By Theorem 5.10, if $A_N(x) \le q$, then x contains powers $x_i^{a_i}$ with all $a_i \ge 2$ and all $|x_i|$ distinct and nonzero such that (5.3) holds:

$$n + 1 - m - \sum_{i=1}^{m}(a_i - 2)|x_i| \le 2q.$$

Applying this with $q = A_N(x)$,

$$n + 1 - m - \sum_{i=1}^{m}(a_i - 2)|x_i| \le 2A_N(x). \tag{5.5}$$

Let $S_i = (a_i - 1)|x_i|$ and $S = \sum_{i=1}^{m} S_i$. Using $|x_i| \ge 1$ and (5.5), we have

1 This part is inspired by an argument in [63].

$$n+1-S \le n+1-S-m+\sum_{i=1}^{m}|x_i| \le 2A_N(x). \tag{5.6}$$

Using $a_i \ge 2$ and the observation that *if m many distinct positive integers $|x_i|$ are all bounded by $\lceil d\log n\rceil$, then it follows that $m \le \lceil d\log n\rceil$ (Lemma 5.13)*, we have

$$\left\{x : \max_{i=1}^{m} S_i \le \lceil d\log n\rceil\right\} \subseteq \left\{x : \max_{i=1}^{m}|x_i| \le \lceil d\log n\rceil\right\} \subseteq \{x : m \le \lfloor d\log n\rfloor\}. \tag{5.7}$$

By (5.4) (since S_i is the length of a continued run in x), we have

$$\mathbb{P}_n\left(\max_{i=1}^{m} S_i \le \lceil d\log n\rceil\right) \ge 1-\epsilon_{n,d}. \tag{5.8}$$

Using $S \le m\max_{i=1}^{m} S_i$, (5.7), and (5.8),

$$\begin{aligned}\mathbb{P}_n(S \le (\lceil d\log n\rceil)^2) &\ge \mathbb{P}_n\left(m\max_i S_i \le (\lceil d\log n\rceil)^2\right)\\ &\ge \mathbb{P}_n\left(\max_{i=1}^{m} S_i \le \lceil d\log n\rceil\right) \ge 1-\epsilon_{n,d}.\end{aligned}$$

So by (5.6),

$$\mathbb{P}_n\left(A_N(x) \ge \frac{n+1}{2} - \frac{1}{2}(\lceil d\log n\rceil)^2\right) \ge 1-\epsilon_{n,d}. \tag{5.9}$$

Letting $n \to \infty$ completes the proof. □

Remark 5.15. Note that this means that Jordon and Moser's sequence with $I(T) = 0$ must have infinitely many n for which $I \upharpoonright n$ contains a segment with lookback amount $d\log_2 n$ for $d > 2$.

5.3 Open problems

Let $\mathcal{C}_2$ be the set of maximally complex binary words satisfying $A_N(x) = \lfloor |x|/2\rfloor + 1$.

For an infinite word $\mathbf{x}$, let $\mathbf{x} \upharpoonright n$ be its prefix of length n.

Let $\mathcal{H}_2$ be the set of infinite words $\mathbf{x}$ with $\mathbf{x} \upharpoonright n \in \mathcal{C}_2$ for all n.

Question n is more likely to have the answer "No" than Question $n+1$ in the following.

Question 5.1. Is there an automatic $\mathbf{x} \in \mathcal{H}_2$? (The Thue–Morse word is not, but almost.)

Question 5.2. Is there a computable $\mathbf{x} \in \mathcal{H}_2$? (Probably yes by bounding how far to look.)

Question 5.3. Is $\mathcal{H}_2$ nonempty?

We conjecture that the answer to Question 5.3 is yes, as the growth rate seems to be $> 2^{n/2}$:

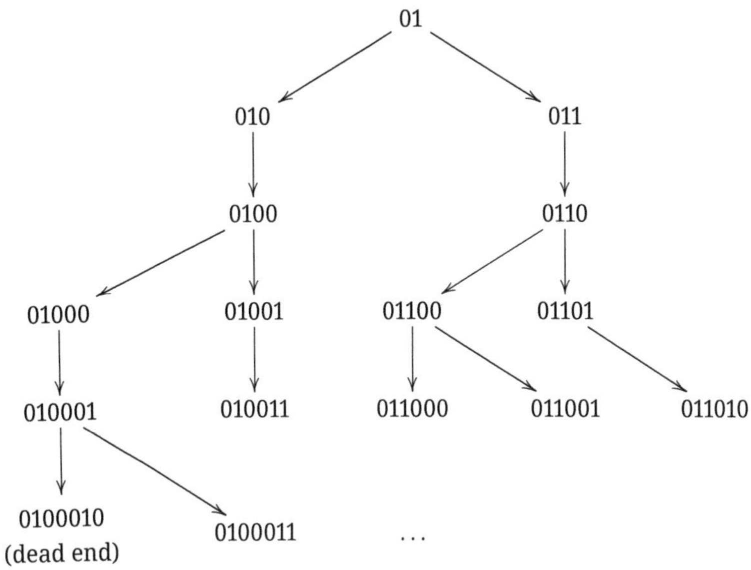

Question 5.4. Is $\mathcal{C}_2$ an infinite set?

Question 5.5. Does $\mathcal{C}_2$ have a member of length 42 or more?

We have confirmed that $\mathcal{C}_2$ has a member of length 41. The word

$$01001100011101100011010101111100101111111$$

of length 41 was generated by the *Rule 75* elementary cellular automaton (a vertical middle segment, skipping the first 9 bits). It has hereditarily maximal nondeterministic automatic complexity. It was found by starting with the leftmost such word of length 27 and using the *Online Encyclopedia of Integer Sequences* [61]. The verification took more than 10 hours.

Rule 75 works like this:

$$111 \to 0$$
$$110 \to 1 \quad 1 \cdot 64$$
$$101 \to 0$$
$$100 \to 0$$
$$011 \to 1 \quad 1 \cdot 8$$

$$\begin{aligned} 010 &\to 0 \\ 001 &\to 1 \quad 1\cdot 2 \\ 000 &\to 1 \quad 1\cdot 1 \\ &\qquad \textstyle\sum = 75 \end{aligned}$$

It generates the length 41 word as follows:

0	0	0	0	1	0	0	0	0	0
1	1	1	1	0	0	1	1	1	1
0	0	0	1	0	1	1	0	0	0
1	1	1	0	0	1	1	0	1	1
0	0	1	0	1	1	1	0	1	0
1	1	0	0	1	0	1	0	0	

5.4 Exercises

Exercise 5.1. Verify that for a word $x = x_0 \dots x_4 \in \{0, 1\}^5$, we have Lookback$(2, 1, 1, x)$ iff $x_1 = x_2$ (Example 5.12).

A complete solution to Exercise 5.1 is available at the following URL.

https://github.com/bjoernkjoshanssen/ac-exercises/

6 Conditional automatic complexity

In this chapter we show that metrics analogous to the Jaccard distance and the normalized information distance can be defined based on conditional nondeterministic automatic complexity A_N. Our work continues the work of Shannon (1950) on entropy metrics and Gács (1974) on symmetry of information among others.

Shallit and Wang (2001) defined the automatic complexity of a word w as, somewhat roughly speaking, the minimum number of states of a finite automaton that accepts w and no other word of length $|w|$. This definition may sound a bit artificial, as it is not clear the length of w is involved in defining the complexity of w. In this chapter we shall see how *conditional* automatic complexity neatly resolves this issue.

6.1 Basics

Definition 6.1 (Track of two words). Let Γ and Δ be alphabets. Let $n \in \mathbb{N}$, $x \in \Gamma^n$, and $y \in \Delta^n$. When no confusion with binomial coefficients is likely, we let $\binom{a}{b} = (a, b) \in \Sigma \times \Delta$. The *track* of x and y, $x\#y \in (\Gamma \times \Delta)^n$, is defined to be the word

$$\binom{x_0}{y_0}\binom{x_1}{y_1}\cdots\binom{x_{n-1}}{y_{n-1}},$$

which we may also denote as $\binom{x}{y}$.

Definition 6.2 (Projections of a word). Let Γ and Δ be alphabets. Let $n \in \mathbb{N}$, $x \in \Gamma^n$, and $y \in \Delta^n$. The projections π_1 and π_2 are defined by $\pi_1(x\#y) = x$, $\pi_2(x\#y) = y$.

Note that $x\#y$ can be thought of as a parametrized curve $i \mapsto (x_i, y_i)$. The symbol # reminds us of the two "tracks" corresponding to x and y. This use of the word "track" can be found in [67].

Theorem 6.3. *There exist words x, y with $A_N(x\#y) \not\leq A_N(x) + A_N(y)$.*

Proof. Let $x = (010)^4$ and $y = (01)^6$. Then we can check that $A_N(x\#y) = 6$, $A_N(x) = 3$, and $A_N(y) = 2$. □

Theorem 6.4. *For all words x, y, we have $\max\{A_N(x), A_N(y)\} \leq A_N(x\#y)$. There exist words x, y with $\max\{A^-(x), A^-(y)\} \not\leq A^-(x\#y)$.*

Proof. Let x be a word of some length n with $A^-(x) > n/2 + 1$. An example can be found among the maximum-length sequences for linear feedback shift registers, as observed in Theorem 3.37. Let y be a rainbow word (Definition 1.5) of the same length. Whenever y is a rainbow word, so is $x\#y$. Therefore, $A^-(x\#y) \leq n/2 + 1 < A^-(x)$. □

https://doi.org/10.1515/9783110774870-006

Definition 6.5. Let Γ and Δ be alphabets. Let $n \in \mathbb{N}$ and $x \in \Gamma^n, y \in \Delta^n$. The *conditional (nondeterministic) automatic complexity of x given y*, $A_N(x \mid y)$, is the minimum number of states of an NFA over $\Gamma \times \Delta$ such that Item i and Item ii hold.

(i) Let m be the number of accepting walks of length $n = |x| = |y|$ for which the word w read on the walk satisfies $\pi_1(w) = y$. Then $m = 1$.

(ii) Let w be the word in Item i. Then $\pi_2(w) = x$.

The conditional complexity $A(x \mid y)$ must be defined in terms of a unique sequence of edges rather than a unique sequence of states, since we cannot assume that there is only one edge from a given state q to given state q'.

Theorem 6.6. *We have $A_N(x\#y) \le A_N(x \mid y) \cdot A_N(y)$. In relativized form, $A_N(x \mid z) \le A_N(y \mid z) \cdot A_N(x \mid y\#z)$.*

Proof. We describe the unrelativized form only. We use a certain product of NFAs.[1] Let two NFAs

$$M_1 = (Q_1, \Gamma \times \Delta, \delta_1, q_{0,1}, F_1), \quad M_2 = (Q_2, \Gamma, \delta_2, q_{0,2}, F_2)$$

be given. The product is $M_1 \times_1 M_2 = (Q_1 \times Q_2, \Delta, \delta, (q_{0,1}, q_{0,2}), F_1 \times F_2)$, where $\delta((q, q'), a) \ni (r, r')$ if $\delta_1(q, (b, a)) \ni r$ and $\delta_2(q', b) \ni r'$ for some b.

(We can also form $M_1 \times_2 M_2$, where $\delta((q, q'), (b, a)) \ni (r, r')$ if $\delta_1(q, (b, a)) \ni r$ and $\delta_2(q', b) \ni r'$.)

For a walk w, let word(w) be the word read on the labels of w. Consider an accepting walk w from (q_1, q_2) to (r_1, r_2). By the definition of the start and final states of $M_1 \times_1 M_2$, the projection $\pi_1(w)$ is also accepting. Hence, by *the $A_N(y)$ witness assumption*, $\pi_1(w)$ is the only accepting walk of its length and word$(\pi_1(w)) = y$. Since word$(\pi_1(w)) = y$, by *the $A_M(x \mid y)$ witness assumption* w is the unique walk with word$(\pi_1(w)) = y$ and word$(\pi_2(w)) = x$. Thus, the accepted word is $x\#y$, and w is the unique accepting walk of its length. □

Theorem 6.6 is not optimal in the following sense.

Theorem 6.7. *There exist x and y with $A_N(x\#y) \ne A_N(x \mid y) \cdot A_N(y)$.*

Proof. Let $y = \texttt{0001}$ and $x = \texttt{0123}$. It is enough to note that $A_N(x\#y) = 3$, $A_N(y) = 2$, and $A_N(x \mid y)$ is an integer. □

However, Theorem 6.6 is optimal in the sense that there is a class of word pairs for which it cannot be improved: let $y = (\texttt{012345})^k$ for some large k, and let $x = (\texttt{0123})^l$, where $4l = 6k$, so that $|x| = |y|$. We have $A_N(x\#y) = \text{lcm}(4, 6) = 12$, $A_N(y) = 6$, and $A_N(x \mid y) = 2$ as witnessed by the NFA:

1 This construction may well have appeared elsewhere, but we are not aware of it.

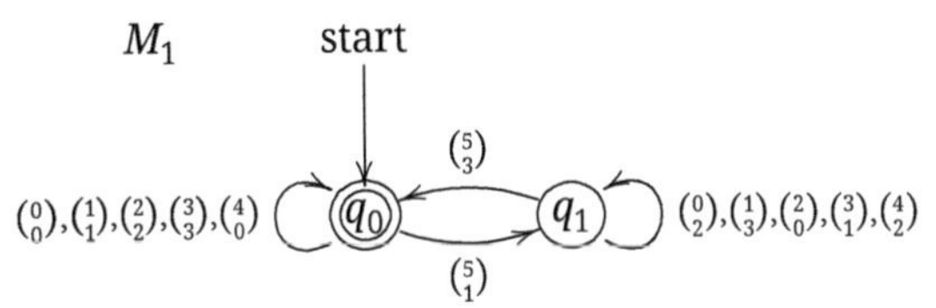

The product of this and a cyclic M_2 automaton for y is another cyclic automaton:

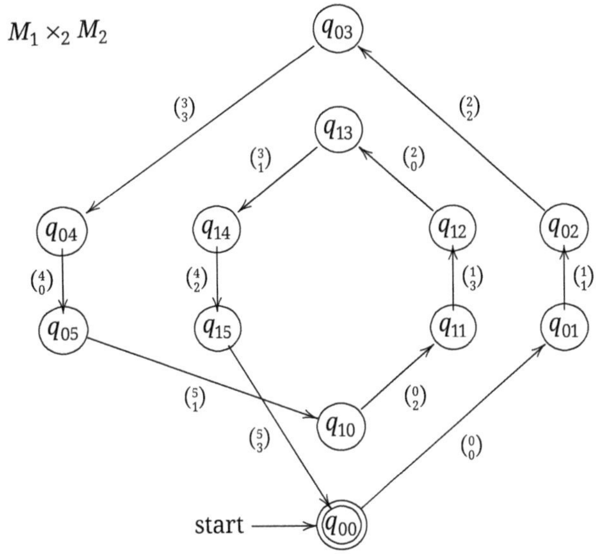

A word x induces an equivalence relation $i \sim_x j \iff x_i = x_j$.

Theorem 6.8. *If $\sim_x$ refines $\sim_y$, then $A_N(y \mid x) = 1$.*
If y is a constant word, then $A_N(x \mid y) = A_N(x)$.

An equivalent characterization is that $A(x \mid y)$ is the minimum number of states of an NFA that accepts y on only one walk (but may accept other words of the same length), such that the equivalence relation induced by the sequence of labeled edges used refines $\sim_x$.

6.2 Bearing on the unique vs. exact problem

A central problem in automatic complexity is whether $A_{Ne} = A_{Nu}$ (Question 1.3). Moving to conditional complexity sheds new light.

Definition 6.9. A *sparse witness* for $A_{Ne}(x \mid y)$ is an NFA M that witnesses the value of $A_{Ne}(x \mid y)$, with the additional properties that:
(i) if any edge is removed from M, then it is no longer a witness of $A_{Ne}(x \mid y)$; and
(ii) M has fewer or an equal number of edges as some witness M_1 of $A_{Nu}(x \mid y)$.

Note that if $A_{Ne}(x \mid y) = A_{Nu}(x \mid y)$, then any witness for $A_{Nu}(x \mid y)$ is a sparse witness for $A_{Ne}(x \mid y)$. The converse fails.

Theorem 6.10. *There exist binary words x, y such that there is a sparse witness for $A_{Ne}(x \mid y)$ that is not an $A_{Nu}(x \mid y)$ witness.*

Proof. We will display slightly more than promised: an $A_{Ne}(x \mid y)$ witness that has *strictly fewer* edges than some $A_{Nu}(x \mid y)$ witness for the same x, y. Consider $x = 0000110$, $y = 0010100$, and the state sequences 00111200 and 01111200. They both generate the same NFA:

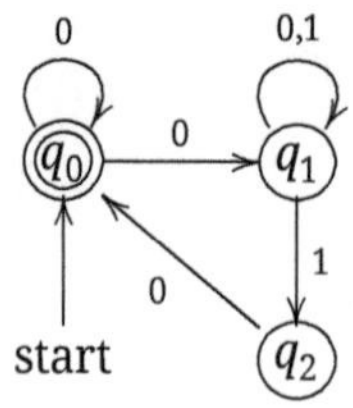

They are sparse witnesses and have only six edges, whereas an A_{Nu} witness with seven edges is the state sequence 01200210:

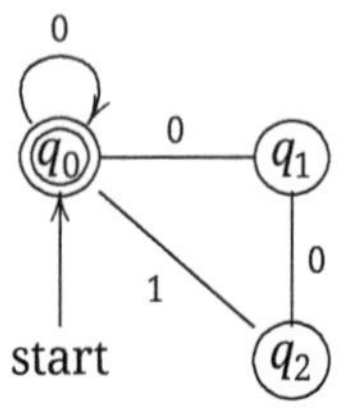

(with y labels)

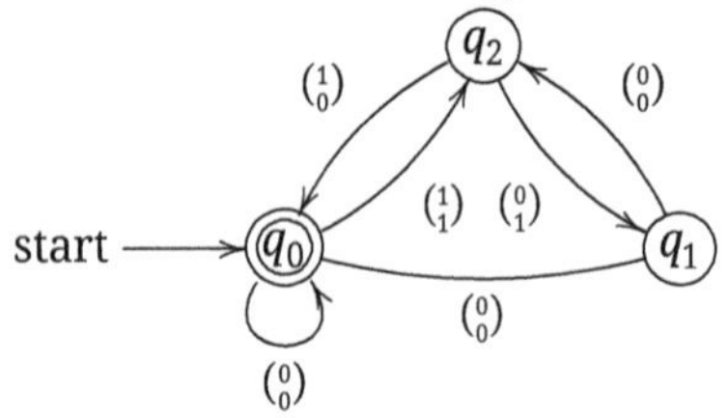

(with $\binom{y}{x}$ labels)

Here, the arrowless edges represent two edges with the same label in opposite directions. □

6.3 A Jaccard distance metric

Let us briefly recall the definitions of metric and pseudometric spaces.

Definition 6.11. Let X be a set. A function $d : X \times X \to \mathbb{R}_{\geq 0}$ is a *pseudometric* if it is commutative, satisfies the triangle inequality, and satisfies $d(x, x) = 0$ for all $x \in X$. If in addition $d(x, y) = 0 \implies x = y$, then d is a *metric*.

The underlying space for our metrics will be $a^n/\mathrm{Sym}(a)$, where a is an alphabet, $n \in \mathbb{N}$, and $\mathrm{Sym}(a)$ is the symmetric group of all permutations of a. Our metrics arise from pseudometrics on a^n. If a is finite we can assume $a = [a] = \{0, 1, \ldots, a-1\}$ for some $a \in \mathbb{N}$ and choose maxshort sequences as representatives. In the case of a binary alphabet $\{0,1\}$, this set of representatives is simply $0\,\{0,1\}^{n-1}$.

Define $J^{\mathrm{num}} : \{(x,y) \in \{0,1\}^* \times \{0,1\}^* : |x| = |y|\} \to \mathbb{R}$ by

$$J^{\mathrm{num}}(x,y) = \log(A_N(x \mid y) A_N(y \mid x)). \tag{6.1}$$

The base of the logarithm chosen is not important, but it is sometimes convenient to let $\log = \log_2$.

Theorem 6.12. *The function J^{num} is a metric on the set $0\{0,1\}^{n-1}$ for any $n \geq 1$.*

Proof. We have $J^{\mathrm{num}}(x,y) = 0$ iff x and y are isomorphic under some permutation of Σ. If we restrict our attention to binary words of the form $0z$, we get $J^{\mathrm{num}}(x,y) = 0 \iff x = y$. And $J^{\mathrm{num}}(x,y) = J^{\mathrm{num}}(y,x)$ is immediate. Theorem 6.6 implies

$$A_N(x \mid y) \leq A_N(z \mid y) \cdot A_N(x \mid z),$$

and hence

$$\begin{aligned} J^{\mathrm{num}}(x,y) &= \log_2(A_N(x \mid y) A_N(y \mid x)) \\ &\leq \log_2(A_N(x \mid z) A_N(z \mid y) A_N(y \mid z) A_N(z \mid x)) \\ &= J^{\mathrm{num}}(x,z) + J^{\mathrm{num}}(y,z). \end{aligned}$$

Hence, the triangle inequality holds. □

The argument in Theorem 6.12 has predecessors: for instance, the simple inequality

$$|A \setminus C| \leq |A \setminus B| + |B \setminus C|$$

was used in the analysis of the Jaccard distance (see Kjos-Hanssen, 2022 [43]). Horibe (1973) [29] gives the following argument, which is credited to a conference talk by Shannon (1950) later published in 1953 [70] (Shannon writes that it is "readily shown" but does not give the argument). Let $H(x \mid y)$ denote the entropy of x given y. Then

$$\begin{aligned} H(x \mid z) &\leq H(x,y \mid z) = H(x \mid y,z) + H(y \mid z) \\ &\leq H(x \mid y) + H(y \mid z). \end{aligned}$$

Similarly, Gács, Tromp, and Vitányi [23] showed

$$K(x \mid y^*) \leq^+ K(x,z \mid y^*) \leq^+ K(z \mid y^*) + K(x \mid z^*).$$

Here, K is the prefix-free Kolmogorov complexity and $K(x \mid y)$ its conditional version; y^* is a minimal-length program for y, so $K(y) = |y^*|$; and $a \leq^+ b$ means $a \leq b + O(1)$. The heavy lifting was already done in 1974 by Gács [22], who showed "symmetry of information," $K(x,y) =^+ K(x) + K(y \mid x^*)$. Symmetry of information does not exactly hold in our setting.

Theorem 6.13. *There exist words x, y with $A_N(x \mid y)A_N(y) \neq A_N(y \mid x)A_N(x)$.*

Proof. Let $x = 0001$, $y = 0011$. Then it would imply $2 \cdot 3 = 2 \cdot 2$. □

Theorem 6.14. *The following function $J^{\mathrm{num}}_{\max}$ is a metric on $0\{0,1\}^{n-1}$:*

$$\begin{aligned} J^{\mathrm{num}}_{\max}(x,y) &= \log_2 \max\{A_N(x \mid y), A_N(y \mid x)\} \\ &= \max\{\log_2 A_N(x \mid y), \log_2 A_N(y \mid x)\}. \end{aligned}$$

Proof. To show the triangle inequality, applying Exercise 6.1,

$$\max\{A(x \mid y), A_N(y \mid x)\} \leq \max\{A_N(x \mid z), A_N(z \mid x)\} \cdot \max\{A_N(y \mid z), A_N(z \mid y)\}.$$ □

We note that $J^{\mathrm{num}}_{\max}$ is an automatic complexity version of the *max complexity* $\max\{K(x \mid y), K(y \mid x)\}$ from [53, Definition 8.3.1].

Lemma 6.15 ([43, Lemma 5]). *Let $d(x,y)$ be a metric and let $a(x,y)$ be a nonnegative symmetric function. If $a(x,z) \leq a(x,y) + d(y,z)$ for all x, y, z, then $d'(x,y) = \frac{d(x,y)}{a(x,y)+d(x,y)}$, with $d'(x,y) = 0$ if $d(x,y) = 0$ is a metric.*

Theorem 6.16. *The following Jaccard distance type function is a metric on $0\{0,1\}^n$ (with the convention $0/0 = 0$):*

$$J(x,y) = \frac{\log(A_N(x \mid y)A_N(y \mid x))}{\log(A_N(x \mid y)A_N(y \mid x)A_N(x)A_N(y)) - \log(A_N(x\#y))}.$$

Proof. It suffices to prove the triangle inequality. We apply Lemma 6.15. Namely, let $d(x,y) = \log_2(A_N(x \mid y)A_N(y \mid x))$ and let $a(x,y) = \log_2(\frac{A_N(x)A_N(y)}{A_N(x\#y)})$. Then we must show $a(x,z) \leq a(x,y) + d(y,z)$, which is equivalent to (writing $A = A_N$ temporarily)

$$\begin{aligned} A(x\#y)A(x)A(z) &\leq A(x)A(y)A(y \mid z)A(z \mid y)A(x\#z), \\ A(x\#y)A(z) &\leq A(y)A(y \mid z)A(z \mid y)A(x\#z). \end{aligned}$$

Since $A(z) \leq A(z \mid y)A(y)$, it suffices to show

$$A(x\#y) \leq A(y \mid z)A(x\#z).$$

This is shown as follows:

$$A(x\#y) \leq A\big(y\#(x\#z)\big) \leq A(y \mid x\#z)A(x\#z) \leq A(y \mid z)A(x\#z).$$ □

One advantage of J is that it does not depend on the base of the logarithm chosen.

Lemma 6.17. *For the special case where $y = 0^n$, a constant word, and $x \neq y$, we have $J(x,y) = 1$.*

Proof. We compute

$$J(x,y) = \log(A_N(x))/\log(A_N(x)A_N(x)/A_N(x)) = 1. \qquad \square$$

If this metric is very close to the discrete metric, it is of course not very interesting. This is fortunately not the case.

Theorem 6.18. *There exist words $x, y \in 0\{0,1\}^{n-1}$ with $0 < J(x,y) < 1/2$, i. e.,*

$$A_N(x)A_N(y) \not\leq A_N(x \mid y)A_N(y \mid x)A_N(x\#y).$$

Proof. Let $x = u0, y = u1$, where $u = 0000100$. Then $J(u0, u1) = 0.46$. $\square$

Theorem 6.19. *Let $x_1, x_2 \in 0\{0,1\}^{n-1}$, $n \leq 12$, with $x_1 \neq x_2$. Let $S = \{x_1, x_2\}$ and let $A = \{001, 010, 011\}$. Then the following are equivalent:*

1. $J(x_1, x_2) = 1$;
2. *either*
   (a) $0^n \in S$, *or*
   (b) $n \geq 10$ *and* $S = \{(01)^{n/2}, a^{n/3}\}$ *for some* $a \in A$.

Proof. This is done by computerized search, so we merely give some remarks on the simplifications making the computation feasible. If there is an example of length 9 or more, then by Theorem 2.2, the equation $A_N(x\#y) = A_N(x)A_N(y)$ must be of the form $a = a \cdot 1, a \leq 5$ (already ruled out) or $4 = 2 \cdot 2$. So it is enough to check words of complexity 2. At length 10 we also have to include the possible equation $6 = 2 \cdot 3$, so we consider all words of complexity 3 or less. $\square$

Lemma 6.20. *If a is a rainbow word and $k \in \mathbb{N}$, then a^k is $(k+1)$-power-free.*

Proof. If $w = a^k$ contains a $(k+1)$-power u, then let a be the first symbol in u. Then a appears at least $k+1$ times in w. However, there are $|a|$ distinct symbols in w and each appears k times. $\square$

Proposition 6.21. *For each nonempty rainbow word a and each $k \in \mathbb{N}$ with $k \geq |a| - 1$, we have $A_N(a^k) = |a|$.*

Proof. Let $a = |a| \geq 1$, $k \geq a - 1$, and $w = a^k$. Then by Lemma 6.20 and Theorem 2.6,

$$A_N(w) \geq \frac{|w|+1}{k+1} = \frac{ka+1}{k+1} = a - \frac{a-1}{k+1} \geq a - \frac{a-1}{a} > a - 1.$$

Since $A_N(w)$ is an integer, $A_N(w) \geq a$. The other direction just uses a single cycle. $\square$

From Proposition 6.21 we have an infinite family of examples with $J(x,y) = 1$ as in Theorem 6.19. Namely, x and y can be powers of rainbow words of relatively prime length.

Lemma 6.22. *If α and β are rainbow words of lengths a and b, then*

$$(\alpha^{\mathrm{lcm}(a,b)/a})\#(\beta^{\mathrm{lcm}(a,b)/b})$$

is a rainbow word.

Proof. Let this word $w = w_1 \dots w_n$, $w_i \in \Sigma$. If $w_i = w_j$, then the first and second coordinates of w_i and w_j are equal, so i is congruent to j mod a and mod b. Hence, i is congruent to j mod $\mathrm{lcm}(a,b)$, so $i = j$. □

The case $k = 2$, $\alpha = 01$, $\beta = 012$ exemplifies Theorem 6.23. There $a = 2$, $b = 3$, $u = 010101$, $v = 012012$, $x = 010101010101$, and $y = 012012012012$.

Theorem 6.23. *Let α and β be rainbow words of relatively prime lengths $a = |\alpha|$ and $b = |\beta|$. Let $x = u^k$, $y = v^k$, where $k \in \mathbb{N}$, $k \geq 2$, $u = \alpha^b$, and $v = \beta^a$. Then $J(x,y) = 1$.*

Proof. It suffices to show $A_N(x\#y) = A_N(x)A_N(y)$. By Lemma 6.22, $u\#v$ is a rainbow word, and $x\#y = (u\#v)^k$, so by Proposition 6.24,

$$A_N(x\#y) = |u\#v| = |u| = ab = A_N(x)A_N(y).$$

□

In Proposition 6.24, the condition $k \geq |\alpha| - 1$ in Proposition 6.21 is strengthened to simply $k \geq 2$ (but of course not to $k \geq 1$).

Proposition 6.21 still gives nonredundant information in the case where $2 > |\alpha| - 1$, i. e., $\alpha \in \{1, 2\}$ and $k \in \{0, 1\}$.

Proposition 6.24. *For each nonempty rainbow word α and $k \geq 2$, we have $A_N(\alpha^k) = |\alpha|$.*

Proof. First assume $k = 2$. Theorem 5.10 implies that if a square of a rainbow word has complexity at most q, then $n/2 \leq q$.

Now let $k > 2$. Since α^2 is a prefix of α^k, we have $A_N(\alpha^k) \geq A_N(\alpha^2) = |\alpha|$. □

We may wonder whether as long as α is primitive, or at least if α has maximal A_N-complexity (achieving Hyde's bound in Theorem 2.2), without necessarily being a rainbow word, Proposition 6.24 still holds, i. e., $A_N(\alpha^2) = |\alpha|$. But this fails.

Definition 6.25. A word w has *emergent simplicity* if $A_N(w)$ is maximal, but $A_N(w^2) < |w|$.

Proposition 6.26. *There exists a word having emergent simplicity. The minimal length of such a word in $\{0,1\}^*$ is 7.*

Proof. Let $w = 0001000$. Then w is maximally complex, but $A_N(w^2) = 6$. The only other example of length 7 is 0010100. In fact, $A^-((0010100)^2) = 6$ as well. □

Hence, not all maximal-complexity words are like 001, 010, and 011 in Theorem 6.19. That is, in general, maximal-complexity words cannot be used to get instances of $J(x,y) = 1$.

We next show that this emergent simplicity can only go so far in Theorem 6.29.

Definition 6.27. A word $\tilde{w}$ is a cyclic shift (or a conjugate) of another word w if there are words a, b with $w = ab$, $\tilde{w} = ba$.

Lemma 6.28 ([67, Theorem 2.4.2]). *A cyclic shift of a primitive word is primitive.*

Theorem 6.29. *If $A_N(w^{|w|}) < |w|$, then w is not primitive (and hence does not have maximal A_N-complexity).*

Proof. If $A_N(w^{|w|}) < |w|$, then $A_N(w^{|w|}) \le |w| - 1$, and hence, since $|w^{|w|}| = |w|^2$ and $k - 1 \not\geq (k^2 + 1)/(k + 1)$ for all k, $A_N(w^{|w|}) \not\geq \frac{|w|^2+1}{|w|+1}$. Therefore, by Theorem 2.6, $w^{|w|}$ contains a $(|w| + 1)$th power $u^{|w|+1}$. Thus, $(|w| + 1)|u| \le |w|^2$, so $|u| < |w|$ (*). Since $w^{|w|}$ also contains $u^{|w|}$, there is a cyclic shift $\tilde{w}$ of w such that $u^{|w|} = \tilde{w}^{|u|}$. Then letting $g = \gcd(|w|, |u|)$, we have $u^{|w|/g} = \tilde{w}^{|u|/g}$ as well, and now the exponents are relatively prime. By Theorem 1.14, there is a word a with $u = a^{|u|/g}$ and $\tilde{w} = a^{|w|/g}$. If $|w|/g > 1$, then $\tilde{w}$ is nonprimitive and hence so is w by Lemma 6.28. If $|w|/g = 1$, then $|w|$ divides $|u|$, which is a contradiction to (*). □

Theorem 6.30. *Let $x, y \in 0\{0,1\}^{n-1}$. Then $A_N(x \mid y) = 1$ iff $x = y$ or $x = 0^n$.*

Proof. Suppose $x \ne 0^n$ and M is a one-state NFA such that there is only one labeled walk on which there exists z with M accepting $y\#z$ (which we can also write $\binom{y}{z}$), and on that one walk, z is x. M has at most four edges, with labels from among $\binom{0}{0}$, $\binom{0}{1}$, $\binom{1}{0}$, $\binom{1}{1}$.

Since $x = 0u$ and $y = 0v$ both start with 0, the walk must start with an edge labeled $\binom{0}{0}$. Then there is no edge labeled $\binom{0}{1}$, or else M would accept both $y\#x$ and $y\#(1u)$.

Since $x \ne 0^n$, x contains a 1, so there is an edge labeled $\binom{0}{1}$ or $\binom{1}{1}$. So there is an edge labeled $\binom{1}{1}$.

Then there is no edge labeled $\binom{1}{0}$, or else two different z's would occur. (Let $x = 0^k1w$; then M would accept both $y\#x$ and $y\#(0^k0w)$.) So M is just the one-state NFA with two edges labeled $\binom{0}{0}$ and $\binom{1}{1}$ only. Consequently, $x = y$. □

This J deems x and y to be "disjoint" when $A_N(x\#y) = A_N(x)A_N(y)$, which for example happens when x and y are high powers of short words of relatively prime length.

6.4 A normalized information distance metric

The following metric $J_{\max}$, more analogous to the normalized information distance [52] than the Jaccard distance, seems better than J above in the following way: if $A_N(x \mid y) = A_N(x)$ and $A_N(y \mid x) = A_N(y)$, then x and y are of no help in compressing each other. This should mean that their distance is maximal ($J_{\max}(x,y) = 1$). This argument can be compared to that made by Li et al. [52]. The condition we get from J, that $J(x,y) = 1$ when $A(x\#y) = A_N(x)A_N(y)$, seems relatively unmotivated in comparison.

Definition 6.31. Let x, y be words of length $n \in \mathbb{N}$. We define

$$J_{\max}(x,y) = \frac{\log \max\{A(x \mid y), A(y \mid x)\}}{\log \max\{A(x), A(y)\}}.$$

Theorem 6.32. *The function $J_{\max}$ is a metric on the set $0\{0,1\}^{n-1}$.*

Proof. The nontrivial part is the triangle inequality. Let

$$a_{xy} = \log \max\{A(x), A(y)\} - \log \max\{A(x \mid y), A(y \mid x)\}.$$

Then by Lemma 6.15 it suffices to show that $a_{xz} \le a_{xy} + d_{yz}$. In other words,

$$\begin{aligned} &\log \max\{A(x), A(z)\} - \log \max\{A(x \mid z), A(z \mid x)\} \\ &\le \log \max\{A(x), A(y)\} - \log \max\{A(x \mid y), A(y \mid x)\} + \log \max\{A(y), A(z)\}. \end{aligned}$$

Equivalently, we must show that

$$\begin{aligned} &\max\{A(x \mid y), A(y \mid x)\} \max\{A(x), A(z)\} \\ &\le \max\{A(x \mid z), A(z \mid x)\} \max\{A(y), A(z)\} \max\{A(x), A(y)\}. \end{aligned}$$

There are now two cases.

Case 1: $A(x) \le A(z)$. Then in fact

$$\begin{aligned} &\max\{A(x \mid y), A(y \mid x)\} \max\{A(x), A(z)\} \\ &\le \max\{A(y), A(z)\} \max\{A(x), A(y)\}. \end{aligned}$$

Case 2: $A(z) < A(x)$. Then we use $A(x \mid y) \le A(x) \le A(x \mid z)A(z)$ and $A(y \mid x) \le A(y)$. □

For this metric the case $J(x,y) = 1$ is perhaps not much easier to understand. We already have examples where $A(y \mid x) = A(y)$ even though x and y are both nontrivial, such as $x, y \in \{001, 010, 011\}$. We can ask whether this metric embeds in Euclidean space, but it fails already at $n = 4$.

The distributions of $q = A_{Nu}(x \mid y)$ for $n \le 10$ are shown in Table 6.1.

Conclusion and future work

We have seen that while automatic complexity A_N is quite different from Kolmogorov complexity K, surprisingly we can obtain metrics similar to those for K using $\log A_N$ and a conditional version of A_N. In fact, these metrics are genuine metrics (not just up to some accuracy) and are computable. We also saw that the conditional version of A_N

Table 6.1: Counts for $A_N(0 :: x \mid 0 :: y), x, y \in \{0,1\}^{n-1}$, $n \geq 1$, and $A_N(\varepsilon \mid \varepsilon)$ $(n = 0)$. Modes are indicated in [brackets]. For example, the table entry for $(n,q) = (2,2)$ is 1 since the only instance of $x, y \in 0\{0,1\}$ with $A_N(x \mid y) = 2$ is $A_N(01 \mid 00)$.

$n \setminus q$	1	2	3	4	5	6
0	[1]					
1	[1]					
2	[3]	1				
3	7	[9]				
4	15	[45]	4			
5	31	[197]	28			
6	63	[755]	191	15		
7	127	[2299]	1561	109		
8	255	5905	[9604]	571	49	
9	511	14005	[47416]	3205	399	
10	1023	31439	[206342]	21066	2102	172

sheds more light upon the problem of distinguishing $A_N = A_{Nu}$ and its word-counting version A_{Ne}. In the future, we hope to characterize when $J = 1$ and $J_{\max} = 1$ and determine whether A_N, J, and $J_{\max}$ are efficiently computable.

6.5 Open problems

Question 6.1. We do not know if sparse witnesses can ever be found in the unconditional case $A_{Ne}(x)$. For example, for $x = 01110$ the state sequence 011220 is a nonsparse witness, and there are no sparse witnesses for any $|x| \leq 8$.

Question 6.2. Is the following problem decidable in polynomial time? *Given x and y, determine whether $J(x,y) = 1$.*

6.6 Exercises

Exercise 6.1. Let $a, b, c, a', b, 'c'$ be natural numbers. Verify that:
(i) if $a \leq bc$ and $a' \leq b'c'$, then $\max\{a, a'\} \leq \max\{b, b'\} \max\{c, c'\}$,
(ii) if $a \leq bc$ and $a' \leq b'c'$, then $aa' \leq (bb')(cc')$, and
(iii) if $a \leq b + c$ and $a' \leq b' + c'$, then $a + a' \leq (b + b') + (c + c')$.

Exercise 6.1 is trivial, but when formalizing Item i it is possible to take a wrong turn. A complete solution to Exercise 6.1 is available at the following URL.

https://github.com/bjoernkjoshanssen/ac-exercises/

7 Logical depth and automatic complexity

7.1 Introduction

Logical depth was defined by Chaitin in 1977 [11] and further studied by Bennett in 1986. It is essentially the time it takes to verify a witness of Kolmogorov complexity. We demonstrate that for automatic complexity, logical depth arises as the difficulty of determining the unique solvability of certain knapsack problems.

A word is deep not so much if it has no simple description but rather if its simplest description is hard to understand and verify. For example, the factorization

$$2001 = 3 \times 23 \times 29$$

is laborious to find, but relatively simple to verify. In that sense, it is not deep. In this chapter we uncover a quite concrete instance of this idea of logical depth. To do so we replace Turing machines by finite automata. In the case of infinite sequences, finite-state depth has been studied by Jordon and Moser [34], but we focus on the concrete world of finite words.

We replace Kolmogorov complexity with *automatic complexity*. The automatic complexity $A(w)$ of a word w was studied by Shallit and Wang (2001) [69]. It is the minimum number of states of a DFA with certain properties and can be defined in a couple of inequivalent variants.

Now, an NFA with a minimum number of edges accepting the word w and no other word of the same length must of course contain a walk on which w is accepted that *visits every edge* of M. It is thus natural to require w to be the only word of length $|w|$ accepted by M on an edge-covering walk, thus obtaining the edge-covering complexity $E_{Nc}(w)$.

Definition 7.1. Let w be a word. The *edge-covering nondeterministic automatic complexity* $E_{Nc}(w)$ is the minimum number of edges of an NFA M that accepts w on an edge-covering walk and has only one accepting edge-covering walk of length $|w|$.

The *edge-covering incomplete deterministic automatic complexity* $E_c(w)$ is the minimum number of edges of an (incomplete) DFA M that accepts w on an edge-covering walk and has only one accepting edge-covering walk of length $|w|$.

Note that there is no total DFA version of Definition 7.1. For instance the word 01 would have no witnessing DFA.

For E_{Nc} we shall obtain a hardness result. It is of the form "does there exist a word w_M for which this NFA M witnesses its $E_{Nc}(w_M)$?" That is in accordance with Bennett's logical depth: given a short description we do not expect to know what it is a description of, until after the long ("deep") computation has finished. We have not yet been able to get a similar result for the other variants of automatic complexity. In that sense, E_{Nc} is chosen so as to make our argument work. Nevertheless, it is a natural modification of automatic complexity.

https://doi.org/10.1515/9783110774870-007

Theorem 7.2. *We have $E_N \neq E_{Nc}$.*

Proof. For a short-word simple word separation of E_N and E_{Nc} we proceed as follows. For the word $w = (01)^3 1(001)^2$ we have $A_N(w) = 6$, $E_N(w) = 8$, both using the state sequence 01212123453450 but $E_{Nc}(w) = 6$ because we can use the state sequence 01010102342342. Both use the equation $2x + 3y = 7$ in different ways.

Let $w = (01)^2 1(001)^2$, which has length 11. We have $A_N(w) = 6$, $E_N(w) = 7$, but $E_{Nc}(w) \leq 6$ using the state sequence 010102342342.

A shorter example, but in a larger alphabet is 0122210 with the state sequence 01222201. It has $E_N = 5$ and $E_{Nc} = 4$. □

We also have $A_N \not\geq E_{Nc}$ by considering the word 012, with $A_N = 2$ and $E_{Nc} = 3$.

Now that we have introduced several notions of automatic complexity, it is time to consider their logical depth and computational complexity. Our associated decision problems come in two flavors depending on how we encode graphs.

Definition 7.3. *Ordinary graph notation* is defined as follows. The set of vertices $\{0, \ldots, q-1\}$ is given by an integer $q \geq 1$ written in binary. Edges can be specified explicitly, by declaring

$$(i, j) \in E,$$

where i, j are written in binary.

Cyclic graph notation extends ordinary graph notation by also allowing edges to be specified by declaring a cycle on an interval $[a, b]$, by the invocation

$$(a, b) \in C.$$

The interpretation is that for $a \leq i < b$, $(i, i+1) \in E$ and $(b, a) \in E$.

Cyclic graph notation can be exponentially more concise than ordinary graph notation. For example, to specify a four-cycle, ordinary notation would look like

$$(00, 01) \in E, \quad (01, 10) \in E, \quad (10, 11) \in E, \quad (11, 00) \in E, \tag{7.1}$$

whereas in cyclic graph notation it would be simply $(00, 11) \in C$.

An *unlabeled NFA* is an NFA where all edge labels have been removed. The information regarding initial and final states is retained. Similarly, a *cursive NFA* is obtained from a cursive digraph (Definition 7.9 below).

Definition 7.4. CYCECAL (Cycle Notation for Edge-Covering Edge-Counting Automatic Complexity Witness at given Length) is the following problem:
- Instance: An unlabeled NFA M, described in cyclic graph notation, and a target length n in binary notation.
- Question: Is there a word w of length n for which some labeling of M is a witness to $E_{Nc}(w)$?

Definition 7.5. ORDECAL (Ordinary Notation for Edge-Covering Edge-Counting Cursive Automatic Complexity Witness at given Length) is the following problem.
- Instance: an unlabeled cursive NFA M, described in ordinary graph notation, and a target length n, given in binary notation.
- Question: is there a word w of length n for which some labeling of M is a witness to $E_{Nc}(w)$?

Theorem 7.6. ORDECAL $\in$ P.

Proof. This reduces to the unique solvability of a knapsack problem, which is weakly polynomial time. Crucially, the parameters in the knapsack problem are given in unary by the NFA. □

If we permit arbitrary, not just cursive, NFAs in the definition of ORDECAL, we do not know whether the problem remains in P. While the determination of whether M has a unique accepting path of length n may be doable in polynomial time as shown for DFAs by Shallit and Wang [69, Theorem 2], we also have to check for edge-covering.

As for CYCECAL, we will show that the complexity is as high as that of UNIQUE SAT.

Definition 7.7. SAT is the following decision problem:
- Instance: A Boolean formula φ.
- Question: Is it true that φ has at least one satisfying truth assignment?

UNIQUE SAT is the following decision problem:
- Instance: A Boolean formula φ.
- Question: Is it true that φ has exactly one satisfying truth assignment?

Blass and Gurevich [8] were interested in the complexity of UNIQUE SAT relative to many-one polynomial time reductions and pointed out the following folklore result.

Lemma 7.8 (Blass and Gurevich). *UNIQUE SAT is coNP-hard.*

Proof. The set of unsatisfiable Boolean formulas, which is well known to be coNP-complete, is reducible to UNIQUE SAT by assigning to each Boolean formula $\varphi(x_1, \dots, x_n)$ the formula $\psi(x_0, \dots, x_n)$ defined as

$$(x_0 \wedge x_1 \wedge \cdots \wedge x_n) \vee (\neg x_0 \wedge \varphi(x_1, \ldots, x_n)).$$

Indeed, suppose φ is unsatisfiable. Then ψ has the unique satisfying assignment $v_0(x_i) = \top$ for each $0 \le i \le n$. On the other hand, if φ is satisfiable, let w be a satisfying assignment of φ. Then ψ has at least two satisfying assignments, namely, v_0 above and v_1 defined by $v_1(x_0) = \bot$, $v_1(x_i) = w(x_i)$ for $i \ge 1$. □

It is also easy to see that UNIQUE SAT belongs to the class Δ_2^P of problems solvable in polynomial time by an algorithm with an NP oracle. The oracle is used to learn whether the given formula has at least one satisfying truth assignment and whether it has at least two.

7.1.1 Cycles as gadgets

To encode knapsack problems into digraphs we use a cycle as a gadget corresponding to an item "price" or "weight" in the knapsack, leading us to a notion of a "cursive" digraph.

Definition 7.9. A *cursive* digraph $G = \mathscr{C}(c_1, \ldots, c_k)$ is a graph of the following form: we have k directed cycle graphs G_i of cardinality c_i, with selected vertices $a_i \in G_i$, $1 \le i \le k$. The graph G is formed from the disjoint union $\bigcup_{i=1}^k G_i$ by adding an edge e_i from a_i to a_{i+1}, for each $1 \le i < k$.

The use of the word "cursive" here is hopefully suggested visually by (7.2):

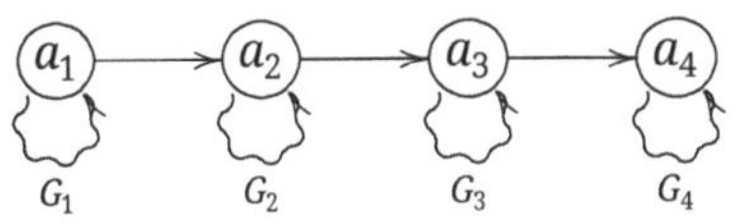

(7.2)

To prove hardness of CYCECAL, we will be obtaining NFAs from digraphs by letting each edge be labeled by itself. Thus, if there is an edge e from q_1 to q_2, then the transition function δ satisfies $\{q_2\} = \delta(q_1, e)$. This way, any uniquely accepted word will be a "cursive word," defined below.

Definition 7.10. Let $\sigma_{a,n}$ be the word over an alphabet $\{e_n \mid n \in \mathbb{N}\}$ defined by $\sigma_{a,0} = \varepsilon$ (the empty word) and $\sigma_{a,n+1} = \sigma_{a,n} e_{a+n}$. Let us say that a *cursive word* is one of the form $\prod_{i=1}^n \sigma_{a_i,b_i}^{c_i}$, where each $a_{i+1} = a_i + b_i$.

For example,

$$\sigma_{0,16} = e_0 e_1 e_2 e_3 e_4 e_5 e_6 e_7 e_8 e_9 e_{10} e_{11} e_{12} e_{13} e_{14} e_{15},$$
$$\sigma_{16,1} = e_{16},$$
$$\sigma_{17,12} = e_{17} e_{18} e_{19} e_{20} e_{21} e_{22} e_{23} e_{24} e_{25} e_{26} e_{27} e_{28},$$

$$\sigma_{29,1} = e_{29},$$
$$\sigma_{30,9} = e_{30}e_{31}e_{32}e_{33}e_{34}e_{35}e_{36}e_{37}e_{38}.$$

A cursive digraph that has a unique walk from a vertex q_0 to a vertex q_f of length n is uniquely associated with a cursive word, as illustrated in Theorem 7.11.

7.1.2 A deep example

To demonstrate the idea of logical depth for automatic complexity, we start with an equation in several variables with a unique solution:

$$16x + 12y + 9z = 83. \tag{7.3}$$

It is easy (although a bit time consuming) to verify that the unique solution over $\mathbb{N}$ is $(x, y, z) = (2, 2, 3)$. This fact has a consequence for automatic complexity, expressed by Theorem 7.11.

Theorem 7.11. *The cursive word*

$$x = \sigma_{0,16}^2\sigma_{16,1}\sigma_{17,12}^2\sigma_{29,1}\sigma_{30,9}^3$$

has $E^-(x) = 39 = \mathrm{wd}(x)$.

Proof. The inequality $E^-(x) \geq \mathrm{wd}(x)$ holds because every symbol of x is read by the NFA and hence requires an edge. For the reverse inequality $E^-(x) \leq 39$ we consider the automaton M in (7.5). There, we have written $e_0, \ldots, e_{38}$ utilizing the Norwegian alphabet as

0123456789ABCDEFGHIJKLMNOPQRSTUVWXÆØÅ.

The unique acceptance of x by M is due to the unique solution in nonnegative integers of the equation

$$16x + 1 + 12y + 1 + 9z = 85 \tag{7.4}$$

and the fact that $|x| = 85$. □

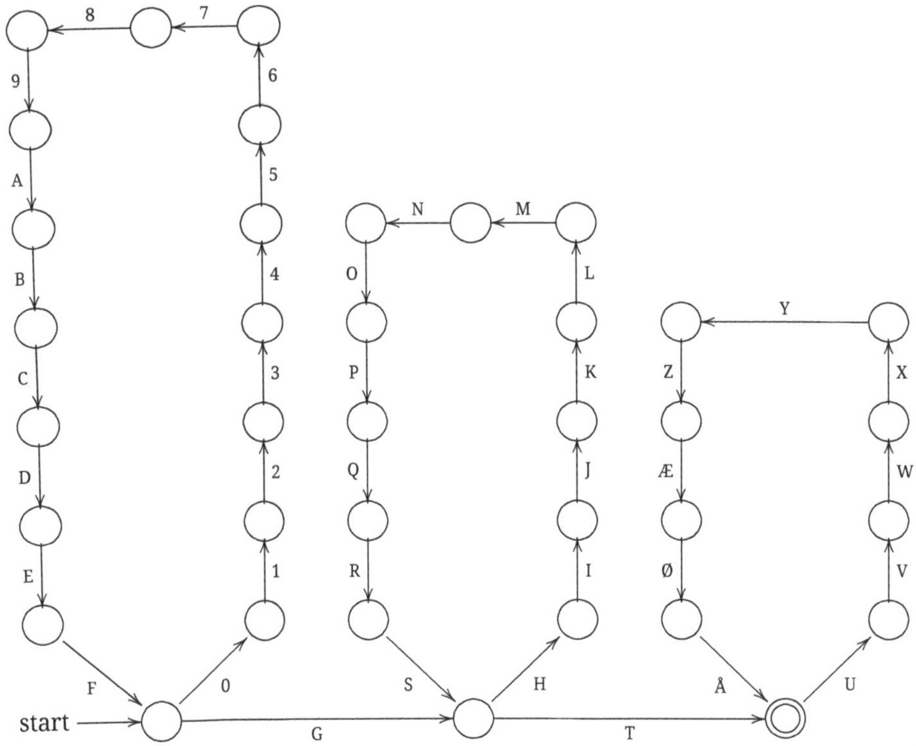

(7.5)

7.1.3 A curious inequality

This subsection can be skipped with no loss of coherence for the reader. However, since our main result (Theorem 7.34) exploits the computational complexity of E_{Nc} for cursive words and A_N is the most important form of automatic complexity, we cannot resist to include the curious Theorem 7.12.

Theorem 7.12. *Let w be a cursive word. Then $A_N(w) \le E_{Nc}(w)$.*

Proof. The obstacle is that witnesses for E_{Nc} are not witnesses for A_N, since there are unwanted accepting walks that exploit "avoidable cycles," which can be traversed zero times. To fix this we make these cycles no longer avoidable by adding an extra state for each such cycle, without having the number of states exceed the number of transitions. The NFA in (7.6) is an A_N witness for the length-22 cursive word $(\texttt{012})^3\texttt{3}(\texttt{4567})^3$. If q_7 is removed and q_3 is made the final state, then the NFA is an E_{Nc} witness but not an A_N witness because the cycle of length 4 is now avoidable:

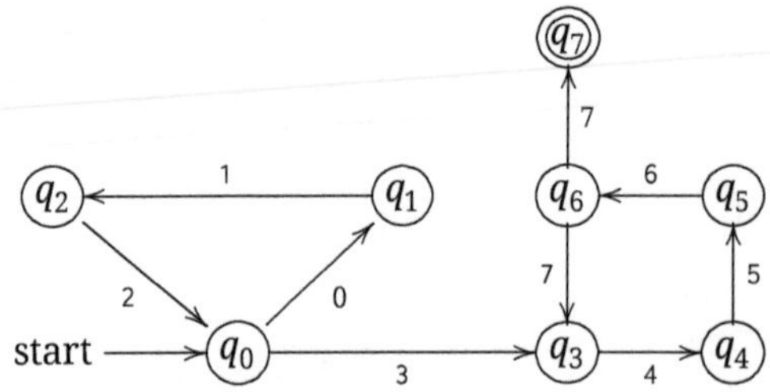

(7.6)

If there was now another cycle, the continuation would look like this:

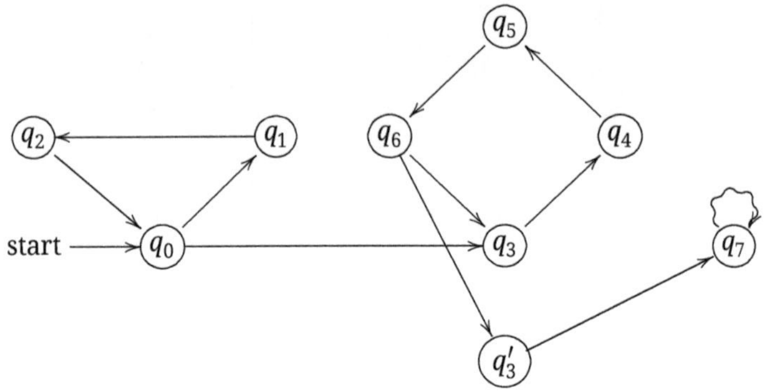

Thus, instead of a direct edge from q_3 to q_7, we insert an extra state q_3' to force any walk to traverse the four-cycle before continuing. □

7.2 Parsimonious reductions

A reduction in which the number of solutions is preserved is *parsimonious* [4]. This is often introduced somewhat informally. To be more precise, we need the notion of a search problem and its associated decision problem.

Definition 7.13. Let A and B be sets and let $R \subseteq A \times B$. The *search problem* (R, A, B) is the set of all partial functions $s : A \to B$ such that for all $x \in A$, if there exists a y with $(x, y) \in R$, then $s(x)$ is defined and $(x, s(x)) \in R$. An element $x \in A$ is called an *instance.* Given an instance x, the set $\{y \mid R(x, y)\}$ is called a *question.*

Informally, the "problem" is to find some such function s and a way to compute it.

Definition 7.14. A *decision problem associated with a search problem* is a pair (A, P), where A is a set, $P = \{x \in A \mid \exists y\, R(x, y)\}$, and $R \subseteq A \times B$ for some set B.

Definition 7.15. Let (A_i, P_i), where $P_i = \{x \in A_i \mid \exists y\, R_i(x, y)\}$ for $i = 1, 2$, be decision problems associated with search problems. A *parsimonious reduction* of (A_1, P_1) to (A_2, P_2) is a function $f : A_1 \times B_1 \to A_2 \times B_2$ whose restriction to R_1 is a bijection between R_1 and R_2.

The UNIQUE version of a decision problem asks whether the decision problem has exactly one solution.

Definition 7.16. Let $Q = (R, A, B)$ be a search problem. The problem UNIQUE Q is the decision problem "given $x \in A$, is there exactly one y in B such that $(x, y) \in R$?"

For computational complexity applications we will also want our reductions to be effective, as in Lemma 7.17 and Lemma 7.19.

Lemma 7.17. *If f is a parsimonious polynomial time computable reduction of the search problem Q_1 to the search problem Q_2, then the same function f is a polynomial time computable many-one reduction of UNIQUE Q_1 to UNIQUE Q_2.*

Proof. This follows immediately from the definitions. □

Definition 7.18. The complexity class US is the class of decision problems solvable by an NP machine such that the answer is "yes" if and only if exactly one computation path accepts.

Lemma 7.19. *If UNIQUE Q_1 is a* US*-hard decision problem and Q_1 is parsimoniously polynomial time many-one reducible to Q_2, then UNIQUE Q_2 is* US*-hard.*

Proof. By Lemma 7.17, UNIQUE Q_1 is polynomial time many-one ($\leq_m^P$) reducible to UNIQUE Q_2. Since the US-hard language is upward closed under $\leq_m^P$, the result follows. □

7.3 Computational complexity

We are ready to start the proof of hardness of CYCECAL. We use a simple modification of INTEGER KNAPSACK that we call POSITIVE KNAPSACK.

Definition 7.20 (Papadimitriou and Steiglitz [60]). INTEGER KNAPSACK is the following problem:
- Instance: Integers c_j, $1 \leq j \leq N$, and K, in binary notation.
- Question: Are there integers $x_j \geq 0$ such that $\sum_{j=1}^{n} c_j x_j = K$?

Definition 7.21. POSITIVE KNAPSACK is the following problem:
- Instance: Integers c_j, $1 \leq j \leq N$, and K (in binary notation).
- Question: Are there integers $x_j > 0$ such that $\sum_{j=1}^{n} c_j x_j = K$?

Lemma 7.22. INTEGER KNAPSACK *is parsimoniously polynomial time many-one reducible to* POSITIVE KNAPSACK.

Proof. Let c_j, $1 \leq j \leq N$, and K form an instance of INTEGER KNAPSACK. Let $c_j' = c_j$ and let $K' = K + \sum_{j=1}^{N} c_j$. Then the INTEGER KNAPSACK instance c_j, K has a solution iff the

POSITIVE KNAPSACK instance c_j', K' has a solution. Moreover, the solutions are in one-to-one correspondence via $x_j \mapsto x_j + 1$. □

Definition 7.23. CLIQUE is the following decision problem:
- Instance: A finite graph G and a positive integer k.
- Question: Is it true that G contains a clique of cardinality k?

Definition 7.24. VERTEX COVER is the following decision problem:
- Instance: A graph $G = (V, E)$ and a positive integer k.
- Question: Is there a subset $V' \subseteq V$ of size at most k such that every edge in the graph is connected to some vertex in V'?

Definition 7.25. SUBSET SUM (called 0–1 KNAPSACK in Papadimitriou and Steiglitz) is the following problem:
- Instance: N nonnegative integers $a_1, \dots, a_N$ and a target sum K.
- Question: Is there is a subset whose sum equals K?

Definition 7.26. CURSIVE COVERING WALK is the following problem:
- Instance: A cursive digraph $G = \mathscr{C}(c_1, \dots, c_k)$, in cyclic graph notation, and an integer n, in binary notation.
- Question: Does G contain an edge-covering walk from a_1 to a_k of length n?

Lemma 7.27. POSITIVE KNAPSACK *is parsimoniously polynomial time many-one reducible to* CURSIVE COVERING WALK.

Proof. Let $c_1, \dots, c_k, K$ be an instance of POSITIVE KNAPSACK. Let G be the cursive digraph $\mathscr{C}(c_1, \dots, c_k)$. Let C_i be the cycle of size c_i (of course, we may assume that each $c_i > 0$) and let $n = K + k - 1$. Then G and n constitute an affirmative instance of CURSIVE COVERING WALK iff $c_1, \dots, c_k, K$ constitute an affirmative instance of POSITIVE KNAPSACK. Indeed, $x_1, \dots, x_k$ is a solution to POSITIVE KNAPSACK iff the walk $C_1^{x_1} e_1 C_2^{x_2} e_2 \dots C_k^{x_k}$ is a solution to CURSIVE COVERING WALK. □

Definition 7.28. A cursive NFA is obtained from a cursive digraph G as follows. The alphabet is the set R of all edges of G. The initial state is a_1 and the final state is a_k. The label of an edge e is e.

Definition 7.29. The edge-covering automatic complexity $E_{Nc}(F)$ of a family $F \subseteq \Sigma^n$ is the minimum number of edges of an NFA M such that F is the set of all words in Σ^n that are read on accepting edge-covering walks of M.

Definition 7.30. CYCECALS (CYCECAL for Sets) is the following problem: given a cursive NFA M, is there a nonempty set of words $F \subseteq R^n$ such that M witnesses that $E_{Nc}(F) = |R|$?

Theorem 7.31. CURSIVE COVERING WALK *is reducible to* CYCECALS.

Proof. Given a cursive digraph G we let M be the cursive NFA of G. If there is an edge-covering walk of length n, then the set of all such walks can be F, as in Definition 7.29, and M witnesses $E_{Nc}(F)$. If there is no edge-covering walk of length n, then no matter how M is obtained by labeling the edges of G, M does not even witness any upper bound on $E_{Nc}(F)$. □

Remark 7.32. Oded Goldreich [25, "Corollaries" before Theorem 6.19] states that "all known reductions among natural NP-complete problems are either parsimonious or can be easily modified to be so." We will not go into the details of this for CLIQUE, VERTEX COVER, and so forth here.

Theorem 7.33. UNIQUE SAT *is* US-*complete.*

Proof. Since SAT is NP-complete, this and the problems in US are of the form UNIQUE Q, where $Q \in$ NP; this follows from Lemma 7.19. □

Theorem 7.34. CYCECAL *is* US-*complete.*

Proof. SAT is polynomial time many-one reducible to CYCECALS by the sequence of parsimonious reductions in Table 7.1. Since CYCECAL is the same thing as UNIQUE CYCECALS, the result follows from Theorem 7.33. □

Example 7.35. Let us inspect the direct reduction of POSITIVE KNAPSACK to CYCECALS. The cursive NFA M associated with a POSITIVE KNAPSACK instance in the case of four

Table 7.1: Parsimonious reductions (see Remark 7.32).

Definition or theorem	Reference
SAT	Definition 7.7
$\leq_m^P$	Well known
3SAT	
$\leq_m^P$	Well known
CLIQUE	Definition 7.23
$\leq_m^P$	Well known
VERTEX COVER	Definition 7.24
$\leq_m^P$	Well known
SUBSET SUM	Definition 7.25
$\leq_m^P$	Papadimitriou and Steiglitz [60]
INTEGER KNAPSACK	Definition 7.20
$\leq_m^P$	Lemma 7.22
POSITIVE KNAPSACK	Definition 7.21
$\leq_m^P$	Lemma 7.27
CURSIVE COVERING WALK	Definition 7.26
$\leq_m^P$	Theorem 7.31
CYCECALS	Definition 7.4

items is shown schematically in (7.7). The length of σ_i is c_i and the knapsack instance is $\{c_0, c_1, c_2, c_3\}$ with a certain target sum K, where the target word length is $K + 3$:

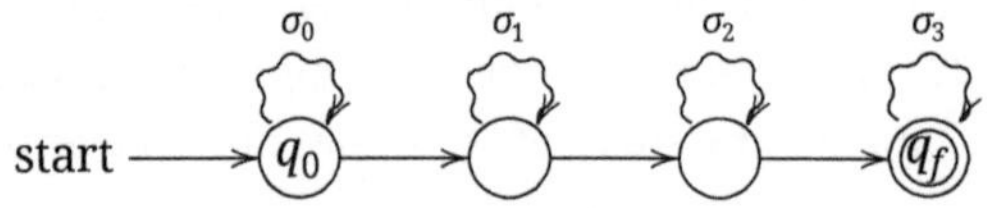

(7.7)

Clearly, M accepts some word of length $K + 3$ on an edge-covering walk iff the POSITIVE KNAPSACK instance $(\{c_0, c_1, c_2, c_3\}, K)$ has a solution. Moreover, there is exactly one word w such that

$$|w| = K + 3 \text{ and } M \text{ accepts } w \text{ on an edge-covering walk}$$

iff there is exactly one tuple (x_0, x_1, x_2, x_3) such that

$$\vec{x} \text{ is a solution to the POSITIVE KNAPSACK instance } (\{c_0, c_1, c_2, c_3\}, K).$$

7.4 Logical depth

Applying parsimoniousness of all reductions, we conclude that the problem CYCECAL is coNP-hard. We remark that this applies equally for determinism and nondeterminism, as all cursive NFAs are (incomplete) deterministic.

The analogy between logical depth for Turing machines and NFAs can be viewed as follows.

Recall HALTING, which is the following problem:

- Instance: A Turing machine T and an input e.
- Question: Does T on input e halt, producing some output x?

In our analogy, the cursive NFA M is analogous to the Turing machine T, the input e is analogous to the length n, and the accepted word w is analogous to the output x. The undecidability of the halting problem is analogous to the coNP-hardness of finding w and determining whether w exists.

Theorem 7.36. *There is a polynomial time algorithm with oracle for* CYCECAL *that computes the cursive word w in the problem* CYCECAL, *if it exists.*

Proof. There is an argument that if SAT is in P, then we can find solutions effectively as well [33]. Similarly, if we take an oracle that answers whether a word w exists as a black box, we can run it on various values for c_1, then for one value of c_1 for which the answer is yes, run it on various values of c_2, etc. □

The depth of x is measured by the time it takes for T to halt and output x. Analogously, the depth of w is measured by the time it takes to obtain w from M and n. Bennett writes that "Depth of a string relative to its length is a particularly useful notion," which can be viewed as presaging our work in a way.

Our results show that one aspect of evaluating a potential automatic complexity witness is US-hard. Let us consider the question whether this is the only difficult aspect of such an evaluation. As argued elsewhere [5], a witnessing NFA for A_N has the digraph structure of a certain tree of cycles, the leaves of which must satisfy a Diophantine equation uniquely. Moreover, there must be no walk that avoids some or all of these leaves. This latter property is in coNP: "there does not exist a solution" (see Example 7.38 below). Since US is not believed to equal coNP, this means that the real difficulty lies in evaluating the Diophantine equations.

In fact, Blass and Gurevich discuss the possibility that US equals Dif^P, which is now usually known as BH_2, and show that the question $\mathsf{US} \stackrel{?}{=} \mathsf{BH}_2$ cannot be resolved by relativizing techniques.

Definition 7.37 (Wechsung [80]). The first two levels of the Boolean hierarchy for NP are given by

$$\begin{aligned} \mathsf{BH}_1 &= \mathsf{NP}, \\ \mathsf{BH}_2 &= \{L_1 \setminus L_2 : L_1, L_2 \in \mathsf{NP}\}. \end{aligned}$$

The complexity classes of interest in this chapter are limited to the following inclusions: $\mathsf{coNP} \subseteq \mathsf{US} \subseteq \mathsf{BH}_2$.

Example 7.38. An NFA witnessing that $A_N(w) \leq 11$, where w is the length-22 word $0^5 1 0^5 1^6 01000$, is shown in (7.8):

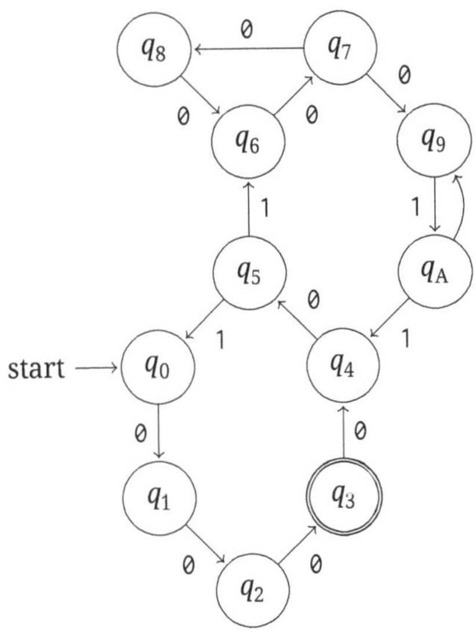

(7.8)

Letting w, x, y, z be the number of times we traverse the four displayed cycles in (7.8), we have the following system of equations and inequalities:

$$\begin{aligned} 3w + 2x + 6y + 6z &= 19, \\ w > 0 \quad &\Longrightarrow \quad y > 0, \\ x > 0 \quad &\Longrightarrow \quad y > 0, \\ y > 0 \quad &\Longrightarrow \quad z > 0. \end{aligned} \tag{7.9}$$

The only solution is $y = z = 1$, $w = 1$, $x = 2$.

In order for a unique walk of length 22 to exist it is clear that $y = z = 1$ would have to be part of the solution. If for example $z = 2$ and $y > 0$, then we could choose to traverse the y cycle after either zero or one trip through the z cycle. If w, x are zero, then there is no solution as 19 is not divisible by 6. That leaves us with the "deep" equation $3w+2x = 7$.

7.4.1 A subword inequality

If a knapsack problem has a unique solution, then so does the problem reduced by removing some items and removing the (secret) amount corresponding to those in the unique solution. This observation has an automatic complexity-theoretic counterpart.

Theorem 7.39. *Let x_1, x_2 be words from disjoint alphabets. Then*

$$E_N(x_1x_2) \geq E_N(x_1) + E_N(x_2) \tag{7.10}$$

and

$$E_{Nc}(x_1x_2) \geq E_{Nc}(x_1) + E_{Nc}(x_2). \tag{7.11}$$

Proof. Let us first prove (7.10). Let M be a witness to $E_N(x_1x_2)$. Let A_i be the set of edges of M used when reading x_i along the unique accepting walk for x_1x_2. We have $E_N(x_i) \leq |A_i|$ since the uniquely accepting walk for x_1x_2 is restricted to a uniquely accepting walk for x_i, if we change the initial state (for x_2) or final state (for x_1) accordingly. Since x_1 and x_2 are words from disjoint alphabets, A_1 and A_2 are disjoint. Hence,

$$E_N(x_1) + E_N(x_2) \leq |A_1| + |A_2| = |A_1 \cup A_2| = E_N(x_1x_2).$$

Now to prove (7.11), we note that a uniquely edge-covering accepting walk y for x_1x_2 is restricted to a uniquely edge-covering accepting walk for x_i. Indeed, suppose that there are two accepting walks y_1, y_2 for x_i that both cover all edges in y that are traversed while reading x_i. Then concatenating either y_i with a walk that covers all edges in y that are traversed while reading x_{3-i} results in an edge-covering walk for M. □

Lemma 7.40. *Let x_1, x_2 be words from disjoint alphabets. If $E_{Nc}(x_1x_2) = \mathrm{wd}(x_1x_2)$, then $E_{Nc}(x_1) = \mathrm{wd}(x_1)$ and $E_{Nc}(x_2) = \mathrm{wd}(x_2)$.*

Proof. We have

$$\begin{aligned}\mathrm{wd}(x_1x_2) &= E_{Nc}(x_1x_2) \quad \text{by assumption}\\ &\geq E_{Nc}(x_1) + E_{Nc(x_2)} \quad \text{by Theorem 7.39}\\ &\geq \mathrm{wd}(x_1) + \mathrm{wd}(x_2)\\ &= \mathrm{wd}(x_1x_2) \quad \text{by Exercise 7.2,}\end{aligned}$$

so $E_{Nc}(x_1) + E_{Nc}(x_2) = \mathrm{wd}(x_1) + \mathrm{wd}(x_2)$. By Exercise 7.1, we are done. □

7.4.2 Future work

Koppel [49] proves a result of the form "sophistication and depth are the same." Here, the *sophistication* of a word w is the minimal complexity of a set S with $w \in S$ such that w is among the most complex members of S. The idea is that neither very simple words nor very complex words are sophisticated, because we can take $S = \{w\}$ or $S = \Sigma^{|w|}$, respectively. It seems that a similar sophistication = depth may occur for automatic complexity. For example, words 0^n1^n are relatively deep but do not have automatic complexity close to the maximum.

Another avenue for future work is the *significance* level of logical depth. Here, a high logical depth occurs at a high significance level if not only is $A_N(w) = k$ (say) with a "deep" witnessing automaton M, but in order to witness the fact that $A_N(w) \leq k + c$ for a moderately sized c we still need "deep" witnesses. We have some anecdotal examples of this phenomenon for automatic complexity and it could be studied in detail.

We end with one such example of high significance for logical depth. The length-23 word $x = (01234)^2 5(678)^4$ has $A_N(x) = 8$. We get $A_N(x) \leq 8$ using the NFA

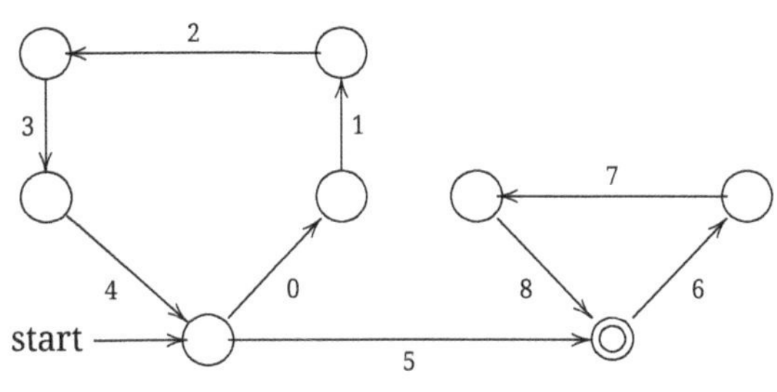

(7.12)

and the equation $5x + 1 + 3y = 23$. If we do not allow NFAs using two-variable equations we cannot even obtain $A_N(x) \leq 11$.

7.5 Exercises

Exercise 7.1. Let a_i, b_i for $i = 1, 2$ be natural numbers and suppose $a_1 + a_2 = b_1 + b_2$ with $a_1 \geq b_1$ and $a_2 \geq b_2$. Then $a_1 = b_1$ and $a_2 = b_2$.

A complete solution to Exercise 7.1 is available at the following URL.

https://github.com/bjoernkjoshanssen/ac-exercises/

Exercise 7.2. Prove the following facts about the width function wd:
1. $\mathrm{wd}(\varepsilon) = 0$;
2. $\mathrm{wd}(a) = 1$ for each symbol a;
3. $\mathrm{wd}(a :: x) = \mathrm{wd}(x)$ if a occurs in x;
4. $\mathrm{wd}(x +\!\!+ y) \leq \mathrm{wd}(x) + \mathrm{wd}(y)$;
5. $\mathrm{wd}(x +\!\!+ y) = \mathrm{wd}(x) + \mathrm{wd}(y)$ if x and y are words from disjoint alphabets;
6. $\mathrm{wd}(x +\!\!+ y) = \mathrm{wd}(y +\!\!+ x)$.

A complete solution to Exercise 7.2 is available at the following URL.

https://github.com/bjoernkjoshanssen/ac-exercises/

Bibliography

[1] Alikhani M, Kjos-Hanssen B, Pakravan A, Saadat B. Pricing complexity options. Algorithmic Finance. 2015; 4(3–4):127–137. Available from: https://doi.org/10.3233/AF-150050.

[2] Althöfer I. Tight lower bounds on the length of word chains. Inf Process Lett. 1990; 34(5):275–276. Available from: https://doi.org/10.1016/0020-0190(90)90136-L.

[3] Arnold A, Brlek S. Optimal word chains for the Thue-Morse word. Inf Comput. 1989; 83(2):140–151. Available from: https://doi.org/10.1016/0890-5401(89)90056-4.

[4] Barbanchon R. On unique graph 3-colorability and parsimonious reductions in the plane. Theor Comput Sci. 2004; 319(1–3):455–482. Available from: https://doi.org/10.1016/j.tcs.2004.02.003.

[5] Beros AA, Kjos-Hanssen B, Yogi DK. Planar digraphs for automatic complexity. In: Theory and applications of models of computation. vol. 11436 of Lecture Notes in Comput. Sci. Cham: Springer; 2019. p. 59–73. Available from: https://doi.org/10.1007/978-3-030-14812-6_5.

[6] Berstel J, Brlek S. On the length of word chains. Inf Process Lett. 1987; 26(1):23–28. Available from: https://doi.org/10.1016/0020-0190(87)90031-7.

[7] Birns S. On the distribution of complexities of finite words, with connections to constructive immunity. University of Hawai'i at Mānoa; 2023.

[8] Blass A, Gurevich Y. On the unique satisfiability problem. Inf Control. 1982; 55(1–3):80–88. Available from: https://doi.org/10.1016/S0019-9958(82)90439-9.

[9] Bousquet-Mélou M. The number of minimal word chains computing the Thue-Morse word. Inf Process Lett. 1992; 44(2):57–64. Available from: https://doi.org/10.1016/0020-0190(92)90186-Y.

[10] Brzozowski JA, Davies S. Most complex deterministic union-free regular languages. In: Descriptional complexity of formal systems. vol. 10952 of Lecture Notes in Comput. Sci. Cham: Springer; 2018. p. 37–48. Available from: https://doi.org/10.1007/978-3-319-94631-3_4.

[11] Chaitin GJ. Algorithmic information theory. IBM J Res Dev. 1977; 21(4):350–359.

[12] Choi JS. Counts of unique periodic binary strings of length n; 2011. http://oeis.org/A152061.

[13] Clark J, Holton DA. A first look at graph theory. Teaneck, NJ: World Scientific Publishing Co., Inc.; 1991. Available from: https://doi.org/10.1142/1280.

[14] Coquand T, Huet G. The calculus of constructions. Inf Comput. 1988; 76(2–3):95–120. Available from: https://doi.org/10.1016/0890-5401(88)90005-3.

[15] Crescenzi P, Goldman D, Papadimitriou C, Piccolboni A, Yannakakis M. On the complexity of protein folding. J Comput Biol. 1998 02;5:423–465.

[16] de Moura LM, Kong S, Avigad J, van Doorn F, von Raumer J. The lean theorem prover (system description). In: Felty AP, Middeldorp A, editors. CADE. vol. 9195 of Lecture Notes in Computer Science. Springer; 2015. p. 378–388. Available from: http://dblp.uni-trier.de/db/conf/cade/cade2015.html#MouraKADR15.

[17] Diwan AA. A new combinatorial complexity measure for languages. Bombay, India: Tata Institute; 1986.

[18] Downey RG, Hirschfeldt DR. Algorithmic randomness and complexity. Theory and Applications of Computability. New York: Springer; 2010. Available from: http://dx.doi.org/10.1007/978-0-387-68441-3.

[19] Erdős P, Selfridge JL. Complete prime subsets of consecutive integers. In: Proceedings of the Manitoba conference on numerical mathematics (Univ. Manitoba, Winnipeg, Man., 1971); 1971. p. 1–14. https://www.renyi.hu/~p_erdos/1971-03.pdf.

[20] Fraenkel AS, Simpson RJ. How many squares must a binary sequence contain? Electron J Comb. 1995; 2:Research Paper 2, approx. 9 pp. (electronic). Available from: http://www.combinatorics.org/Volume_2/Abstracts/v2i1r2.html.

[21] Fujie F, Zhang P. Covering walks in graphs. SpringerBriefs in Mathematics. New York: Springer; 2014. Available from: https://doi.org/10.1007/978-1-4939-0305-4.

[22] Gač P. The symmetry of algorithmic information. Dokl Akad Nauk SSSR. 1974; 218:1265–1267.

https://doi.org/10.1515/9783110774870-008

[23] Gács P, Tromp JT, Vitányi PMB. Algorithmic statistics. IEEE Trans Inf Theory. 2001; 47(6):2443–2463. Available from: https://doi-org.eres.library.manoa.hawaii.edu/10.1109/18.945257.

[24] Gel'fond AO. Sur les nombres qui ont des propriétés additives et multiplicatives données. Acta Arith. 1967/1968; 13:259–265.

[25] Goldreich O. Computational complexity. Cambridge: Cambridge University Press; 2008. A conceptual perspective. Available from: https://doi.org/10.1017/CBO9780511804106.

[26] Greenberg N, Jain S, Ng KM, Schewe S, Stephan FC, Wu G, et al.. Aspects of computation and automata theory with applications. G – Reference, Information and Interdisciplinary Subjects Series. World Scientific Publishing Company Incorporated; 2023. Available from: https://books.google.com/books?id=ZLUC0AEACAAJ.

[27] Halberstam H, Richert HE. Sieve methods. vol. 4 of London Mathematical Society Monographs, London–New York: Academic Press [Harcourt Brace Jovanovich, Publishers]; 1974.

[28] Hopcroft JE, Ullman JD. Introduction to automata theory, languages, and computation. Addison-Wesley Series in Computer Science. Reading, Mass.: Addison-Wesley Publishing Co.; 1979.

[29] Horibe Y. A note on entropy metrics. Inf Control. 1973; 22:403–404.

[30] Hyde K. Nondeterministic finite state complexity. University of Hawaii at Manoa. U.S.A.; 2013.

[31] Hyde KK, Kjos-Hanssen B. Nondeterministic automatic complexity of overlap-free and almost square-free words. Electron J Comb. 2015; 22(3):Paper 3.22, 18.

[32] Iwaniec H. On the error term in the linear sieve. Acta Arith. 1971; 19:1–30. Available from: https://doi.org/10.4064/aa-19-1-1-30.

[33] Jones K. https://cs.stackexchange.com/users/114/kyle-jones. Is finding a solution of a satisfiability problem harder than deciding satisfiability?; URL: https://cs.stackexchange.com/q/29483 (version: 2021-09-12). Computer Science Stack Exchange.

[34] Jordon L, Moser P. On the difference between finite-state and pushdown depth. In: SOFSEM 2020: theory and practice of computer science. vol. 12011 of Lecture Notes in Comput. Sci. Cham: Springer; 2020. p. 187–198. Available from: https://doi.org/10.1007/978-3-030-38919-2_16.

[35] Jordon L, Moser P. Normal sequences with non-maximal automatic complexity. In: Bojanczyk M, Chekuri C, editors. 41st IARCS annual conference on foundations of software technology and theoretical computer science, FSTTCS 2021, December 15–17, 2021, virtual conference. vol. 213 of LIPIcs. Schloss Dagstuhl – Leibniz-Zentrum für Informatik; 2021. p. 47:1–47:16. Available from: https://doi.org/10.4230/LIPIcs.FSTTCS.2021.47.

[36] Kjos-Hanssen B. Kolmogorov structure functions for automatic complexity. Theor Comput Sci. 2015; 607:435–445. Available from: https://doi.org/10.1016/j.tcs.2015.05.052.

[37] Kjos-Hanssen B. On the complexity of automatic complexity. Theory Comput Syst. 2017; 61(4):1427–1439. Available from: https://doi-org.eres.library.manoa.hawaii.edu/10.1007/s00224-017-9795-4.

[38] Kjos-Hanssen B. Superposition as memory: unlocking quantum automatic complexity. In: Unconventional computation and natural computation. vol. 10240 of Lecture Notes in Comput. Sci. Cham: Springer; 2017. p. 160–169. Available from: https://doi.org/10.1007/978-3-319-58187-3_12.

[39] Kjos-Hanssen B. Automatic complexity of shift register sequences. Discrete Math. 2018; 341(9):2409–2417. Available from: https://doi.org/10.1016/j.disc.2018.05.015.

[40] Kjos-Hanssen B. Few paths, fewer words: model selection with automatic structure functions. Exp Math. 2019; 28(1):121–127. Available from: https://doi.org/10.1080/10586458.2017.1368048.

[41] Kjos-Hanssen B. Automatic complexity of Fibonacci and Tribonacci words. Discrete Appl Math. 2021; 289:446–454. Available from: https://doi.org/10.1016/j.dam.2020.10.014.

[42] Kjos-Hanssen B. An incompressibility theorem for automatic complexity. Forum Math Sigma. 2021; 9:Paper No. e62, 7. Available from: https://doi.org/10.1017/fms.2021.58.

[43] Kjos-Hanssen B. Interpolating between the Jaccard distance and an analogue of the normalized information distance. J Log Comput. 2022; 32(8):1611–1623. Available from: https://doi-org.eres.library.manoa.hawaii.edu/10.1093/logcom/exac069.

[44] Kjos-Hanssen B. Conditional automatic complexity and its metrics. In: COCOON 2023. vol. TBA of Lecture Notes in Comput. Sci. Cham: Springer; 2023. p. TBA.
[45] Kjos-Hanssen B. Complexity lookup; http://math.hawaii.edu/wordpress/bjoern/complexity-of-0110100110010110/.
[46] Kjos-Hanssen B. Complexity option game; http://math.hawaii.edu/wordpress/bjoern/complexity-option-game/.
[47] Kolmogorov AN. Three approaches to the definition of the concept "quantity of information". Probl Pereda Inf. 1965; 1(vyp. 1):3–11.
[48] Kolmogorov AN. Three approaches to the quantitative definition of information. Int J Comput Math. 1968; 2:157–168. Available from: https://doi.org/10.1080/00207166808803030.
[49] Koppel M. Complexity, depth, and sophistication. Complex Syst. 1987; 1(6):1087–1091.
[50] Kozen D, Silva A. Practical coinduction. Math Struct Comput Sci. 2017; 27(7):1132–1152. Available from: https://doi.org/10.1017/S0960129515000493.
[51] Krieger D. On critical exponents in fixed points of non-erasing morphisms. Theor Comput Sci. 2007; 376(1–2):70–88. Available from: https://doi.org/10.1016/j.tcs.2007.01.020.
[52] Li M, Chen X, Li X, Ma B, Vitányi PMB. The similarity metric. IEEE Trans Inf Theory. 2004; 50(12):3250–3264. Available from: https://doi.org/10.1109/TIT.2004.838101.
[53] Li M, Vitányi P. An introduction to Kolmogorov complexity and its applications. 2nd ed. Graduate Texts in Computer Science. New York: Springer-Verlag; 1997. Available from: https://doi-org.eres.library.manoa.hawaii.edu/10.1007/978-1-4757-2606-0.
[54] Lucia. https://mathoverflow.net/users/38624/lucia. Unpublished result of Rosser in sieve methods book; URL:https://mathoverflow.net/q/416328 (version: 2022-02-16). MathOverflow.
[55] Lyndon RC. On Burnside's problem. Trans Am Math Soc. 1954; 77:202–215. Available from: https://doi.org/10.2307/1990868.
[56] Lyndon RC, Schützenberger MP. The equation $a^M = b^N c^P$ in a free group. Mich Math J. 1962; 9:289–298.
[57] Morgenbesser JF, Shallit J, Stoll T. Thue-Morse at multiples of an integer. J Number Theory. 2011; 131(8):1498–1512. Available from: http://dx.doi.org/10.1016/j.jnt.2011.02.006.
[58] Morse M, Hedlund GA. Symbolic dynamics II. Sturmian trajectories. Am J Math. 1940; 62:1–42. Available from: https://doi.org/10.2307/2371431.
[59] Nagy B. Union-free regular languages and 1-cycle-free-path-automata. Publ Math (Debr). 2006; 68(1–2):183–197. Available from: https://doi.org/10.5486/pmd.2006.3303.
[60] Papadimitriou CH, Steiglitz K. Combinatorial optimization: algorithms and complexity. USA: Prentice-Hall, Inc.; 1982.
[61] Price R. Middle column of the "Rule 75" elementary cellular automaton starting with a single ON (black) cell; 2016. https://oeis.org/A266895.
[62] Quas A. https://mathoverflow.net/users/11054/anthony-quas. Group action with unique word; URL: https://mathoverflow.net/q/372714 (version: 2020-09-27). MathOverflow.
[63] Quas A. https://mathoverflow.net/users/11054/anthony-quas. Longest runs and concentration of measure; 2016. URL: https://mathoverflow.net/q/247929 (version: 2016-08-21). MathOverflow.
[64] Roth P. A note on word chains and regular languages. Inf Process Lett. 1989; 30(1):15–18. Available from: https://doi.org/10.1016/0020-0190(89)90167-1.
[65] Sakarovitch J. Elements of automata theory. Cambridge: Cambridge University Press; 2009. Translated from the 2003 French original by Reuben Thomas. Available from: https://doi-org.eres.library.manoa.hawaii.edu/10.1017/CBO9781139195218.
[66] Schimmerling E. A course on set theory. Cambridge: Cambridge University Press; 2011. Available from: https://doi.org/10.1017/CBO9780511996351.
[67] Shallit J. A second course in formal languages and automata theory. 1st ed. New York, NY, USA: Cambridge University Press; 2008.

[68] Shallit J, Breitbart Y. Automaticity. I. Properties of a measure of descriptional complexity. J Comput Syst Sci. 1996; 53(1):10–25. Available from: https://doi.org/10.1006/jcss.1996.0046.
[69] Shallit J, Wang MW. Automatic complexity of strings. J Autom Lang Comb. 2001; 6(4):537–554. 2nd Workshop on Descriptional Complexity of Automata, Grammars and Related Structures (London, ON, 2000).
[70] Shannon C. The lattice theory of information. Trans IRE Prof Group Inf Theory. 1953; 1(1):105–107.
[71] Shelton RO, Soni RP. Chains and fixing blocks in irreducible binary sequences. Discrete Math. 1985; 54(1):93–99. Available from: http://dx.doi.org/10.1016/0012-365X(85)90065-2.
[72] Simpson SG. Subsystems of second order arithmetic. 2nd ed. Perspectives in logic. Cambridge: Cambridge University Press; Poughkeepsie, NY: Association for Symbolic Logic; 2009. Available from: https://doi.org/10.1017/CBO9780511581007.
[73] Sipser M. A complexity theoretic approach to randomness. In: Proceedings of the fifteenth annual ACM symposium on theory of computing. STOC '83. New York, NY, USA: ACM; 1983. p. 330–335. Available from: https://dl.acm.org/doi/10.1145/800061.808762.
[74] Sipser M. Introduction to the theory of computation. Cengage Learning; 2012. Available from: https://books.google.com/books?id=1aMKAAAAQBAJ.
[75] Sloane NJA. Number of halving and tripling steps to reach 1 in '3x+1' problem, or −1 if 1 is never reached; 1991. https://oeis.org/A006577.
[76] Sloane NJA. Van Eck's sequence; 2010. https://oeis.org/A181391.
[77] Solomonoff RJ. A formal theory of inductive inference. I. Inf Control. 1964; 7:1–22.
[78] Solomonoff RJ. A formal theory of inductive inference. II. Inf Control. 1964; 7:224–254.
[79] Thue A. Über die gegenseitige Lage gleicher Teile gewisser Zeichenreihen. Norske Vid Skrifter I Mat-Nat Kl, Christiania. 1912; 1:1–67.
[80] Wechsung G. On the Boolean closure of NP. In: Fundamentals of computation theory (Cottbus, 1985). vol. 199 of Lecture Notes in Comput. Sci. Berlin: Springer; 1985. p. 485–493. Available from: http://dx.doi.org/10.1007/BFb0028832.

Index

https://doi.org/10.1515/9783110774870-009

De Gruyter Series in Logic and Its Applications

Volume 11
Isaac Goldbring (Ed.)
Model Theory of Operator Algebras, 2023
ISBN 978-3-11-076821-3, e-ISBN (PDF) 978-3-11-076828-2,
e-ISBN (EPUB) 978-3-11-076833-6

Volume 10
Gabriel Goldberg
The Ultrapower Axiom, 2022
ISBN 978-3-11-071969-7, e-ISBN (PDF) 978-3-11-071973-4,
e-ISBN (EPUB) 978-3-11-071979-6

Volume 9
Merlin Carl
Ordinal Computability. An Introduction to Infinitary Machines, 2019
ISBN 978-3-11-049562-1, e-ISBN (PDF) 978-3-11-049615-4,
e-ISBN (EPUB) 978-3-11-049291-0

Volume 8
Chi Tat Chong, Liang Hu
Recursion Theory, 2015
ISBN 978-3-11-027555-1, e-ISBN (PDF) 978-3-11-027564-3
e-ISBN (EPUB) 978-3-11-038129-0

Volume 7
Itay Neeman
The Determinacy of Long Games, 2008
ISBN 978-3-11-018341-2, e-ISBN 978-3-11-020006-5

Volume 6
Godehard Link (Ed.)
One Hundred Years of Russell's Paradox, 2008
ISBN 978-3-11-017438-0, e-ISBN 978-3-11-019968-0

Volume 5
Martin Zeman
Inner Models and Large Cardinals, 2011
ISBN 978-3-11-016368-1, e-ISBN 978-3-11-085781-8

www.ingramcontent.com/pod-product-compliance
Ingram Content Group UK Ltd.
Pitfield, Milton Keynes, MK11 3LW, UK
UKHW050953290926
15470UKWH00006B/1796

9 783110 774818